KINETIC AND NONSTEADY-STATE EFFECTS IN SUPERCONDUCTORS

KINETIC AND NONSTEADY-STATE EFFECTS IN SUPERCONDUCTORS

B.T. Geilikman and V.Z. Kresin

Translated from Russian by P. Shelnitz
Translation edited by IPST Staff

A HALSTED PRESS BOOK

JOHN WILEY & SONS
New York · Toronto

ISRAEL PROGRAM FOR SCIENTIFIC TRANSLATIONS
Jerusalem · London

Sole distributors for the Western Hemisphere and Japan

HALSTED PRESS, a division of
JOHN WILEY & SONS, INC., NEW YORK

Library of Congress Cataloging in Publication Data

Geilikman, Boris Tov'evich
Kinetic and nonsteady-state effects in superconductors.

Translation of *Kineticheskie i nestat͡sionarnye i͡avlenii͡a v sverkhprovodnikakh.*
"A Halsted Press book."
Bibliography: p.
1. Superconductors. I. Kresin, Vladimir Zinov'evich, joint author. II. Title.
QC612.S8G3813 537.6'23 73-22089
ISBN 0-470-29513-9

Distributors for the U.K., Europe, Africa and the Middle East

JOHN WILEY & SONS, LTD., CHICHESTER

Distributed in the rest of the world by

KETER PUBLISHING HOUSE JERUSALEM LTD.
ISBN 0 7065 1428 9
IPST cat.no. 22110

This book is a translation from Russian of
KINETICHESKIE I NESTATSIONARNYE YAVLENIYA
V SVERKHPROVODNIKAKH
Izdatel'stvo "Nauka"
Moscow, 1972

Printed in Israel

CONTENTS

INTRODUCTION

When a substance passes from the normal to the superconducting state, its physical properties undergo a sharp change. The electromagnetic, thermal and kinetic properties of superconductors differ from those of normal metals, which are described fairly well by the usual quantum solid-state theory. It is shown in the microscopic theory of superconductivity /1—3/ that the main features of the superconducting state are due to pair correlations of electrons and to the associated appearance of a gap in the energy spectrum of superconductors.

This book is devoted to a description of kinetic and nonsteady-state phenomena in superconductors; transfer phenomena in superconductors were discussed briefly in previous surveys of superconductivity theory /4—8/, which appeared both immediately after the advent of the microscopic theory and in recent years.

It is evident that the enormous volume of relevant theoretical and experimental material precludes a uniformly complete discussion of the subject. Our main attention will be concentrated on transfer processes: heat conduction and absorption of sound in the usual superconductors. We believe that the study of these highly important phenomena has now reached a level at which a fairly detailed phenomenological description is feasible. We do not consider questions relating to the tunnel effect in superconductors; these have been discussed fairly adequately in the books /9, 10/, which appeared in the series "Problems of Modern Physics."

The investigation of heat transfer processes in superconductors is quite young. The specific features of these phenomena (see e.g., Shoenberg /11/) inspired the two-fluid model /12/, which furnishes an adequate qualitative description of some of the observed regularities.

In recent years, the modern microscopic theory of superconductivity has been the basis for a fairly complete investigation of heat conduction and sound absorption in superconductors. It is our opinion that the contemporary physics of the superconducting state provides a highly satisfactory description of transfer processes.

The temperature dependence of the various kinetic coefficients proves to be quite different from that described by the formulas fo normal metals. The principal peculiarity of superconductors is th existence of a condensate of Cooper pairs, whose motion is not accompanied by entropy transfer. The number of electron excitations, which play a major role in kinetic phenomena, depends on the temperature, decreasing as the latter falls below the critical temperature T_c and approaching zero as $T \to 0$. This temperature dependence of the number of carriers and the corresponding change in the free path are the source of the unusual pattern of kinetic phenomena in superconductors.

The author is highly indebted to A. G. Shepelev for many useful comments concerning the absorption of sound in superconductors.

Chapter I

THERMAL CONDUCTIVITY OF SUPERCONDUCTORS

1. HEAT CONDUCTION MECHANISMS

As is well known, the heat flux in a metal is composed of two components: the electronic thermal conductivity, $\varkappa_e$, and the lattice thermal conductivity, $\varkappa_p$, so that

$$\varkappa = \varkappa_e + \varkappa_p.$$

The stationary nature of the process of heat conduction is ensured by the presence of several relaxation mechanisms. In the study of the electronic component of the heat flow, the mechanisms are collisions between electrons and impurities (which determine $\varkappa_{ed}$), electron-phonon collisions ($\varkappa_{ep}$) and electron-electron collisions ($\varkappa_{ee}$). The lattice conductivity $\varkappa_p$ is determined by interactions between phonons and electrons ($\varkappa_{pe}$), phonon-phonon collisions ($\varkappa_{pp}$), and scattering of phonons by lattice impurities, crystal boundaries, or crystal defects ($\varkappa_{pd}$). Thus there are six mechanisms for heat conduction.

Accurately calculating the thermal conductivity clearly requires solving the complete kinetic equation which takes into account all the mechanisms. However, as is known from solid-state quantum theory, the thermal resistivity for a given type of carrier is approximately equal to the sum of the thermal resistivities corresponding to the various mechanisms. Therefore,

$$\varkappa_e^{-1} \approx \varkappa_{ed}^{-1} + \varkappa_{ep}^{-1} + \varkappa_{ee}^{-1}, \quad \varkappa_p^{-1} \approx \varkappa_{pe}^{-1} + \varkappa_{pp}^{-1} + \varkappa_{pd}^{-1}.$$

Of course, depending on conditions, the various mechanisms play different roles. In an ordinary metal the electronic component, $\varkappa_e$ makes the primary contribution, while the other component, $\varkappa_p$, usually amounts to only several percent of the total conductivity.

The temperature dependence of the electronic thermal conductivity for a normal metal, $\varkappa_e^n$, is adequately expressed by the equation

$$1/\varkappa_e^n = \alpha T^2 + \beta/T. \tag{1.1}$$

The first term on the right in equation (1.1) represents the interaction of the electrons with the lattice vibrations, and the second term is related to the scattering of electrons by impurities, defects, and boundaries in the crystal. In an ideal crystal, the thermal conductivity would be determined only by the first term in equation (1.1) and would approach infinity as $T \to 0$. However, at sufficiently low temperatures, the electrons begin to be scattered primarily by the inevitable defects and impurities. This process gradually results in a linear decrease of conductivity as T approaches zero. At a certain temperature, whose value is determined by the degree of purity of the specimen, a maximum conductivity is observed in normal metals. As the concentration of the impurities increases, this maximum is naturally shifted toward higher temperatures.

Rewriting (1.1) in the form

$$T/\varkappa_e^n = \alpha T^3 + \beta,$$

we arrive at the conclusion that $T/\varkappa_e^n$ must be linearly dependent on T^3. The ordinate intercept will then equal β, and the slope of the line will determine the value of the coefficient α. Such linear functions are indeed observed experimentally. As expected, when an impurity is introduced, the slope of the straight line does not change. Only a parallel displacement, corresponding to an increase in β, is observed.

The experimentally observed linear dependence of $T/\varkappa^n$ on T^3 indicates that the contribution of the lattice conductivity to the heat flow in a normal metal is small. Only in a very strongly doped metal, which has a small electronic free path and correspondingly low electronic thermal conductivity, does the role of the lattice conductivity increase.

In dielectrics, the lattice conductivity is very high. Its small value in metals is caused by the strong scattering of phonons by free electrons. We will see below that one of the peculiarities of superconductors is that the nature of their heat conduction becomes similar to that of dielectrics in the low temperature range.

Thus, the normal metals are characterized by the fact that the electronic component, $\varkappa_e^n$, plays the major role in their heat conduction.

The electronic thermal conductivity also plays the major role in superconductors for $T \to T_c$ and in pure or slightly doped specimens in the intermediate temperature range. When T approaches zero and $\varkappa_e^s$ falls exponentially (see below), the role of $\varkappa_p^s$ increases substantially even in substances containing a high concentration of impurity. A more detailed discussion of the role of the various mechanisms of heat conduction in superconductors will be given in Sec. 6 where we calculate the thermal conductivities for the various mechanisms. Incidentally, since the basic properties of the superconducting state are associated with a rearrangement of the electron system, the values of $\varkappa_{pp}^{s}$ and $\varkappa_{pd}^{s}$ are described by the same laws as in normal metals.

2. SCATTERING OF ELECTRONS BY IMPURITIES

This mechanism /14/ plays a major role in superconducting specimens containing a small concentration of impurity. At the same time, of course, the temperature range near the absolute zero is excluded, where the number of electron excitations is exponentially low.

Heat flux is defined by the relationship ($\hbar = 1$)

$$Q = -\varkappa \frac{\partial T}{\partial x} = 2\int \varepsilon v_x f \frac{dp}{(2\pi)^3},$$

where f is the perturbed distribution function of the electron excitations in a superconductor which see a temperature gradient and interact with the impurity atoms. To evaluate it, we write the corresponding kinetic equation

$$\frac{\partial f}{\partial x}\frac{\partial \varepsilon}{\partial p_x} - \frac{\partial f}{\partial p_x}\frac{\partial \varepsilon}{\partial x} = -\frac{f - f_0}{\tau}. \tag{1.2}$$

Here $\varepsilon = (\xi^2 + \Delta^2(T))^{1/2}$ is the energy of the electron excitation ($\xi = (p^2 - p_F^2)/2m \approx v_F(p - p_F)$ is the ground-state electron energy reckoned from the Fermi surface, v_F and p_F are the velocity and momentum on the Fermi surface and $\Delta(T)$ is the gap in the energy spectrum); $f_0 = [\exp(\varepsilon/T) + 1]^{-1}$ (the temperature is measured in energy units, $k = 1$).

We substitute the equilibrium function f_0 in place of f on the left-hand side of equation (1.2), which then becomes

$$\frac{\partial f_0}{\partial \varepsilon}\frac{\varepsilon}{T}\frac{\partial \varepsilon}{\partial p_x}\frac{\partial T}{\partial x} = \frac{f - f_0}{\tau}. \tag{1.2'}$$

Note that the temperature dependence of the energy of excitations drops out from the kinetic equation.

The relaxation time τ of the electron excitations interacting with impurities in a superconductor appears on the right in equation (1.2'). To evaluate this relaxation time, we write the second quantization Hamiltonian for the interaction of the ground-state electrons with impurity atoms:

$$H' = \sum_{k,\,k'} V_{kk'}(a^+_{k,\,1/2}a_{k',\,1/2} + a^+_{k,\,-1/2}a_{k',\,-1/2})$$

($a^+_{k,\,1/2}$ and $a_{k,\,-1/2}$ are the amplitudes of the second quantization with momentum k and spin $1/2$ and $-1/2$, corresponding to the creation and destruction of an electron; $V_{kk'}$ is a matrix element).

According to Bogolyubov /2/, the superconducting state is described by Fermi operators α_{k0} and α_{k1}, where

$$\alpha_{k0} = u_k a_{k,\,1/2} - v_k a^+_{-k,\,-1/2}, \quad \alpha_{k1} = u_k a_{-k,\,-1/2} + v_k a^+_{k,\,1/2}, \tag{1.3}$$
$$u^2_k = {}^1/_2(1 + \xi/\varepsilon), \quad v^2_k = {}^1/_2(1 - \xi/\varepsilon).$$

Introducing these operators into H', we find for elastic scattering ($|k| = |k'|$)

$$H' = \sum_{k,\,k'} (\xi/\varepsilon) V_{kk'}(\alpha^+_{k0}\alpha_{k'0} + \alpha^+_{k1}\alpha_{k'1})$$

($\alpha^+_{ki}\alpha_{ki}$ are the number operators of excitations).

Next, expressing the probability of scattering of an electron excitation with the aid of the well-known quantum mechanical formula $W_{ab} = 2\pi\,|H'|^2_{ab}\rho_E$ ($\rho_E = p^2 d\Omega dp/d\varepsilon$ is the statistical weight), we find

$$\tau = \tau_0\varepsilon/|\xi| \tag{1.4}$$

(τ_0 is the energy-independent relaxation time of the ground-state electrons).

It is obvious from equation (1.4) that in a superconductor the free paths of the ground-state electron and the electron excitation should be equal (the ratio of the velocity of an electron in a normal metal, v^n, to the velocity in a superconductor, $v^s = \partial\varepsilon\,\partial p$, is also equal to $\varepsilon/|\xi|$).

Substituting the expression for τ into equation (1.2), we find the sought distribution function, and then calculate the thermal conductivity $\varkappa^s_{ed}$:

$$\varkappa^s_{ed} = \frac{2}{3}\frac{p_F^3\tau_0}{\pi^2 m} F(T), \tag{1.5}$$

$$F(T) = T^{-1}\int\limits_{\infty}^{\Delta} \varepsilon^2 \frac{\partial f_0}{\partial\varepsilon} d\varepsilon = \frac{\Delta^2}{T}\left[\exp\left(\frac{\Delta}{T}\right) + 1\right]^{-1} +$$
$$+ 2T\sum_{s=1}^{\infty} \frac{(-1)^{s+1}}{s^2} \exp\left(-\frac{s\Delta}{T}\right) + 2\Delta\ln\left[1 + \exp\left(-\frac{\Delta}{T}\right)\right].$$

Thus we see from (1.5) that the electronic thermal conductivity is described by a universal function of temperature: $\varkappa_{ed}$ decreases with a decrease in T. The graph of this function, together with experimental data obtained from a study of the conductivity of doped Tl specimens are shown in Figure 1. The theoretical curve fits the experimental data well.

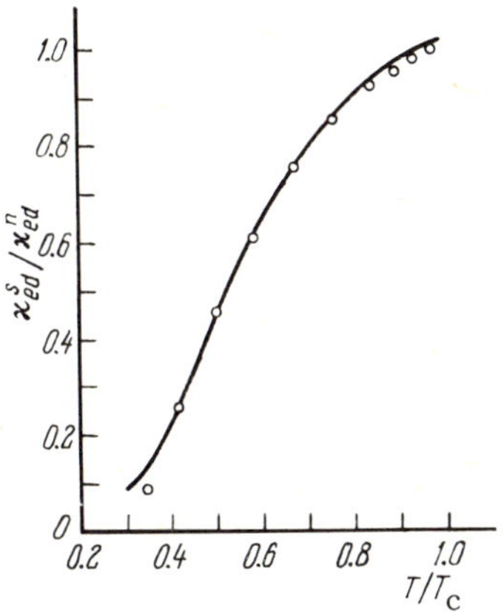

FIGURE 1. Temperature dependence of $\varkappa^s_{ed}$. The experimental data were obtained from a study of the conductivity of Tl /15/. The solid line is theoretical /14/.

If we set $\Delta = 0$ in (1.5), we obtain the linear relationship between $\varkappa^n_{ed}$ and temperature which is derived in the usual theory of metals (see, for example, /13/).

A function for $\varkappa^s_{ed}$ similar to the one presented in Figure 1 has since been obtained independently in /16/. The influence of paramagnetic impurities has been discussed in /131/.

In /17, 18/ the electronic thermal conductivity is calculated using the exact expression for thermal conductivity /19/ (see Sec. 8) and applying the method of Green's functions. As a result, the authors arrive at the same equation, (1.5), which is derived in /14/ with the aid of the kinetic equation.

3. THERMAL CONDUCTIVITY OF PURE SUPERCONDUCTORS

In very pure specimens, for $T \to T_c$, the major factor is $\varkappa_{ep}$, the thermal conductivity associated with the scattering of electrons by phonons.

To study the temperature dependence of $\varkappa_{ep}^{s}$ /20, 21/, we write the kinetic equation:

$$\frac{\partial f}{\partial x}\frac{\partial \varepsilon}{\partial p_x} - \frac{\partial f}{\partial p_x}\frac{\partial \varepsilon}{\partial x} = \left(\frac{\partial f}{\partial t}\right)_{\text{col}}$$

where f is the electron distribution function, which is sought in the form

$$f = f_0 + \frac{\partial f_0}{\partial \varepsilon}\varphi(\varepsilon, \Omega)\frac{\partial T}{\partial x}. \tag{1.6}$$

In order to write the expression for the collision integral in this case, we make the transformation (1.3) to the new Fermi operators in the Hamiltonian for the electron-phonon interaction,

$$H' = \sum_{k, s, q} V_{kk'} a^{+}_{ks} a_{k's} b^{+}_{q} + \text{h. c.} \tag{1.7}$$

(b^{+}_{q} is the creation operator of a phonon with momentum q; $s = {}^1/_2$ and $-{}^1/_2$; $V_{kk'}$ is a matrix element; $k' = k + q$), and the original Hamiltonian becomes

$$H' = \sum_{k, q} V_{kk'}[(u_k u_{k'} - v_k v_{k'})(\alpha^{+}_{k0}\alpha_{k'0} + \alpha^{+}_{k1}\alpha_{k'1}) + \\ + (u_k v_{k'} + u_{k'} v_k)(\alpha^{+}_{k0}\alpha^{+}_{k'1} + \alpha_{k0}\alpha_{k'1})]\, b^{+}_{q} + \text{h. c.} \tag{1.7'}$$

Considering this expression for H', we write the kinetic equation in the form

$$-\frac{\partial f}{\partial \varepsilon}\frac{\partial \varepsilon}{\partial p_x}\frac{\varepsilon}{T}\frac{\partial T}{\partial x} = \sum_{q}\Big[|V|^2\{[f'(1-f)(N+1) - f(1-f')N]\,\delta(\varepsilon' - \varepsilon - \hbar\omega) + \\ + [f'(1-f)N - f(1-f')(N+1)]\,\delta(\varepsilon - \varepsilon' - \hbar\omega)\} \times \\ \times\left(1 + \frac{\xi\xi' - \Delta^2}{\varepsilon\varepsilon'}\right) + |V|^2\,\frac{1}{2}\left(1 - \frac{\xi\xi' - \Delta^2}{\varepsilon\varepsilon'}\right) \times \\ \times [N(1-f)(1-f') - (N+1)ff']\,\delta(\varepsilon + \varepsilon' - \hbar\omega)\Big], \tag{1.8}$$

$$|V|^2 = |V_{kk'}|^2_F.$$

We seek the distribution function in the form of (1.6). Substituting (1.6) into the kinetic equation (1.8), we find

$$f_0^2 e^{\varepsilon/T}\frac{\xi}{T}\frac{p_x}{m} = \int |V|^2 N_0\left(1 + \frac{\xi\xi' - \Delta^2}{\varepsilon\varepsilon'}\right)[\varphi(\varepsilon', \Omega') - \varphi(\varepsilon, \Omega)] \times \\ \times e^{\varepsilon'/T} f_0(\varepsilon) f_0(\varepsilon')\,\delta(\varepsilon' - \varepsilon - \hbar\omega)\, dq + \\ + \int |V|^2 N_0\left(1 + \frac{\xi\xi' - \Delta^2}{\varepsilon\varepsilon'}\right)[\varphi(\varepsilon', \Omega') - \varphi(\varepsilon, \Omega)] \times$$

$$\times e^{\varepsilon/T} f_0(\varepsilon) f_0(\varepsilon') \delta(\varepsilon - \varepsilon' - \hbar\omega)\, d\boldsymbol{q} +$$
$$+ \int |V|^2 N_0 \left(1 - \frac{\xi\xi' - \Delta^2}{\varepsilon\varepsilon'}\right) [\varphi(\varepsilon', \Omega') - \varphi(\varepsilon, \Omega)] \times$$
$$\times e^{\hbar\omega/T} f_0(\varepsilon) f_0(\varepsilon') \delta(\varepsilon' + \varepsilon - \hbar\omega)\, d\boldsymbol{q}, \qquad (1.8')$$
$$N_0 = \left[e^{\frac{\hbar\omega}{T}} - 1\right]^{-1}.$$

We next note that the magnitude of the phonon wave vector $\boldsymbol{q}$ is less than $|\boldsymbol{p}|$. We therefore expand $\varphi(\varepsilon', \Omega')$ in powers of $\boldsymbol{q} = \boldsymbol{p}' - \boldsymbol{p}^*$ ($\boldsymbol{p}^*$ is a vector which has the same direction as the vector $\boldsymbol{p}$ and length $|\boldsymbol{p}'|$). This method for the investigation of the kinetic equation was first developed by Landau and Pomeranchuk /22/. Integrating over the angle ϑ (the polar axis is chosen in the direction of the vector $\boldsymbol{p}$), we obtain

$$f_0^2 e^{\varepsilon/T} \frac{\varepsilon}{T} \frac{\xi}{|\xi|} \frac{p_x}{m} = \int_0^{\infty} N_0 |V|^2 \frac{\varepsilon\varepsilon'}{|\xi||\xi'|} \left(1 + \frac{\xi\xi' - \Delta^2}{\varepsilon\varepsilon'}\right) [\varphi(\varepsilon + \hbar\omega, \Omega) - \varphi(\varepsilon, \Omega)] \times$$
$$\times \frac{e^{\frac{\varepsilon + \hbar\omega}{T}}}{\left(e^{\frac{\varepsilon}{T}} + 1\right)\left(e^{\frac{\varepsilon + \hbar\omega}{T}} + 1\right)} \frac{m}{pq} q^2 dq\, d\varphi +$$
$$+ \int_0^{\varepsilon - \Delta} N_0 |V|^2 \frac{\varepsilon\varepsilon'}{|\xi||\xi'|} \left(1 + \frac{\xi\xi' - \Delta^2}{\varepsilon\varepsilon'}\right) [\varphi(\varepsilon - \hbar\omega, \Omega) - \varphi(\varepsilon, \Omega)] \times$$
$$\times \frac{e^{\varepsilon/T}}{\left(e^{\frac{\varepsilon}{T}} + 1\right)\left(e^{\frac{\varepsilon - \hbar\omega}{T}} + 1\right)} \frac{m}{pq} q^2 dq\, d\varphi +$$
$$+ \int_{\varepsilon + \Delta}^{\infty} N_0 |V|^2 \frac{\varepsilon\varepsilon'}{|\xi||\xi'|} \left(1 - \frac{\xi\xi' - \Delta^2}{\varepsilon\varepsilon'}\right) [\varphi(\hbar\omega - \varepsilon, \Omega) - \varphi(\varepsilon, \Omega)] \times$$
$$\times \frac{e^{\frac{\hbar\omega}{T}}}{\left(e^{\frac{\varepsilon}{T}} + 1\right)\left(e^{\frac{\hbar\omega - \varepsilon}{T}} + 1\right)} \frac{m}{pq} q^2 dq\, d\varphi + \int_0^{\infty} N_0 |V|^2 \frac{\varepsilon\varepsilon'}{|\xi||\xi'|} \left(1 + \frac{\xi\xi' - \Delta^2}{\varepsilon\varepsilon'}\right) \times$$
$$\times \frac{e^{\frac{\varepsilon + \hbar\omega}{T}}}{\left(e^{\frac{\varepsilon}{T}} + 1\right)\left(e^{\frac{\varepsilon + \hbar\omega}{T}} + 1\right)} \frac{m}{pq} \frac{\partial^2 \varphi}{\partial \Omega^2} q_i q_j q^2\, dq\, d\varphi + \int_0^{\varepsilon - \Delta} N_0 |V|^2 \frac{\varepsilon\varepsilon'}{|\xi||\xi'|} \left(1 + \frac{\xi\xi' - \Delta^2}{\varepsilon\varepsilon'}\right) \times$$
$$\times \frac{e^{\frac{\varepsilon}{T}}}{\left(e^{\frac{\varepsilon}{T}} + 1\right)\left(e^{\frac{\varepsilon - \hbar\omega}{T}} + 1\right)} \frac{m}{pq} \frac{\partial^2 \varphi}{\partial \Omega^2} q_i q_j q^2 dq\, d\varphi + \int_{\varepsilon + \Delta}^{\infty} N_0 |V|^2 \frac{\varepsilon\varepsilon'}{|\xi||\xi'|} \left(1 - \frac{\xi\xi' - \Delta^2}{\varepsilon\varepsilon'}\right) \times$$
$$\times \frac{e^{\frac{\hbar\omega}{T}}}{\left(e^{\frac{\varepsilon}{T}} + 1\right)\left(e^{\frac{\hbar\omega - \varepsilon}{T}} + 1\right)} \frac{m}{pq} \frac{\partial^2 \varphi}{\partial \Omega^2} q_i q_j q^2 dq\, d\varphi. \qquad (1.8'')$$

Evaluation of $L = \varepsilon\varepsilon' [1 \pm (\xi\xi' - \Delta^2)/\varepsilon\varepsilon']/|\xi||\xi'|$ shows that $L \approx 2$ when $\xi\xi' > 0$ in the case of scattering (+) and when $\xi\xi' < 0$ in the case of pair creation (-). In all other cases, L is negligible (the error is approximately $\sim \Delta^2 (\hbar\omega)^2/4(\varepsilon\varepsilon' \mp \Delta^2)\,\varepsilon\varepsilon'$). We next integrate both sides of equation (1.8'') over ξ after expanding the left side of (1.8'') in powers of $p_x/p = (p_x/p)_{p_F} + a_x\xi$. The integral in the first term vanishes since the integrand is odd. We now seek a solution in the form

$$\varphi(\varepsilon, \Omega) = \varphi_1(\Omega) + \varphi_2(\varepsilon, \Omega), \qquad \varphi_2 \ll \varphi_1.$$

Integrating over ε with the aid of the substitution $d\xi = \varepsilon\, d\varepsilon/\xi$, we obtain

$$\begin{aligned}
\int_b^\infty f_0^2 e^z a_x z (z^2 - b^2)^{1/2} dz &= |V'|^2 T^2 \int_0^\infty \frac{x^2 dx}{e^x - 1} \int_b^\infty [\varphi_2(z + x, \Omega) - \\
&- \varphi_2(z, \Omega)] \frac{e^{z+x} dz\, d\varphi}{(e^z + 1)(e^{z+x} + 1)} + \\
&+ |V'|^2 T^2 \int_0^\infty \frac{x^2 dx}{e^x - 1} \int_{b+x}^\infty [\varphi_2(z - x, \Omega) - \varphi_2(z, \Omega)] \times \\
&\times \frac{e^z dz\, d\varphi}{(e^z + 1)(e^{z-x} + 1)} + |V'|^2 T^2 \int_{2b}^\infty \frac{x^2 dx}{e^x - 1} \int_b^{x-b} [\varphi_2(x - z, \Omega) - \\
&- \varphi_2(z, \Omega)] \frac{e^x dz\, d\varphi}{(e^z + 1)(e^{x-z} + 1)} + \\
&+ |V'|^2 T^4 \int_0^\infty \frac{x^4 dx}{e^x - 1} \int_b^\infty \frac{e^{z+x} dz}{(e^z + 1)(e^{z-x} + 1)} \frac{\partial^2 \varphi_1}{\partial \Omega^2} + \\
&+ |V'|^2 T^4 \int_0^\infty \frac{x^4 dx}{e^x - 1} \int_{b+x}^\infty \frac{e^z dz}{(e^z + 1)(e^{z-x} + 1)} \frac{\partial^2 \varphi_1}{\partial \Omega^2} + \\
&+ |V'|^2 \frac{T^4}{2} \int_{2b}^\infty \frac{x^4 dx}{e^x - 1} \int_b^{x-b} \frac{e^x d\varphi\, dz}{(e^z + 1)(e^{x-z} + 1)} \frac{\partial^2 \varphi_1}{\partial \Omega^2}.
\end{aligned} \qquad (1.8''')$$

Here $\varepsilon/T = z$, $\hbar\omega/T = x$, $\Delta/T = b$, and $|V'|^2 = |V|^2/q$. The first integrals, which contain $\varphi_2(\varepsilon, \Omega)$, vanish (to prove this, substitute $z + x$ for z in the second integrand and $x - z$ for z in the third). We now find that

$$\varphi = \frac{\alpha(\Omega)}{T^4} \frac{\int_b^\infty f_0^2 e^z z (z^2 - b^2)^{1/2} dz}{\Phi(T)}, \qquad (1.9)$$

$$\Phi(T) = \int_0^\infty \frac{4x^4 dx}{e^x - 1} \int_b^\infty \frac{dz}{(e^z + 1)(e^{-z-x} + 1)} + \int_{2b}^\infty \frac{x^4 dx}{e^x - 1} \int_b^{x-b} \frac{dz}{(e^z + 1)(e^{-z} + e^{-x})}.$$

With the aid of equation (1.8'), it can easily be shown, as in /22/, that φ_2 increases more slowly than φ_1 with decreasing temperature; and thus, in the temperature range of interest, φ_2 is actually much less than φ_1.

Evaluating the integrals in $\Phi(T)$, and subsequently calculating the heat flux Q according to the equation

$$Q = -\varkappa\frac{\partial T}{\partial x} = 2\int \varepsilon v_x f\frac{dp}{(2\pi\hbar)^3},$$

we obtain the following final expression:

$$\varkappa_{ep}^{s} = \frac{\text{const}}{\Phi(T)T^2}\left[b^2\sum_{s=1}^{\infty}(-1)^{s+1}K_2(bs)\right]^2 \tag{1.10}$$

($K_2(bs)$ is the Bessel function of an imaginary argument; $\Phi(T)$ is approximately described by (3.41).

In the temperature range near T_c, where the contribution of the electron-phonon interaction to the heat flux is the most substantial, we find that

$$\varkappa_{ep}^{s} = \varkappa_{ep}^{n}\frac{36}{\pi^4}\frac{\Phi(T_c)}{\Phi(T)}\left[2\sum_{s=1}^{\infty}\frac{(-1)^{s+1}}{s^2}e^{-bs} + 2b\ln(1+e^{-b}) - \frac{b^2}{2(e^b+1)}\right]^2 \tag{1.10'}$$

($\varkappa_{ep}^{n}$ is the thermal conductivity of a normal metal). The corresponding theoretical curve* and experimental data for pure specimens of tin and indium are given in Figure 2.

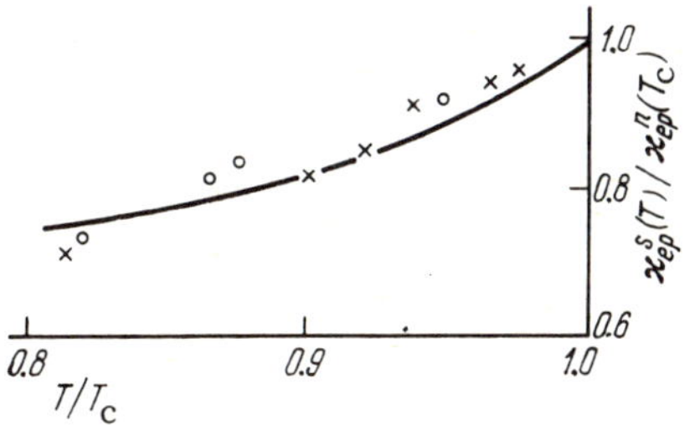

FIGURE 2. $\varkappa_{ep}^{s}$ as a function of the temperature. The experimental points correspond to tin (crosses) /23/ and indium (circles) /24/; the solid curve is theoretical /20/.

* Note that (1.10') is also given in /21/, but there is an error in the corresponding table of numerical values. A detailed derivation of (1.10) and (1.10') is given in /20/, where, in particular, the function $\varkappa_{ep}^{s}(T)$ is accurately tabulated (see Figure 2).

Equation (1.8') can also be solved by another method. The function $\varphi(\varepsilon, \Omega)$ can be expanded into a series in Legendre polynomials. Furthermore, we may restrict ourselves to only the first term of the expansion (see Chapter 3, Sec. 10). The same temperature dependence (1.9) is obtained for the function φ. As a result we arrive at equations (1.10) and (1.10'), which describe the electronic thermal conductivity.

The curve illustrated in Figure 2 is quite different from the one obtained in /16/. In /16/ a variational method was used and a function was obtained in which the electronic thermal conductivity of a superconductor, $\varkappa_{ep}^{s}$, increased with decreasing temperature below T_c. In fact, the decrease in the number of electron excitations, which corresponds to this fall in temperature, leads to a decrease in the electronic thermal conductivity. The decrease in the electronic thermal conductivity is cited in many experimental works.

The contradiction between the theory developed in /16/ and experiment can be resolved without any additional assumptions, such as those introduced in /25—27/. The fact is, as we have already stated, that the calculations were carried out by the variational method, whereas the form of the distribution function adopted in /16/ is markedly different from functions (1.6) and (1.9), obtained by an analytic solution of the kinetic equation.

In sufficiently pure superconductors it is also necessary to consider effects associated with the presence of overlapping bands (for further details, see Sec. 9).

4. LATTICE CONDUCTIVITY

We shall now discuss heat transfer by phonons /31, 32/, which is associated with the scattering of phonons by electron excitations in the superconductor. This mechanism plays a decisive role in the temperature range ($T \gtrsim (0.3—0.5)\, T_c$).

We write the kinetic equation for the phonon distribution function N:

$$-\frac{\partial N}{\partial T}\frac{\partial T}{\partial x}u_0\frac{q_x}{|\boldsymbol{q}|} = \left(\frac{\partial N}{\partial t}\right)_{\text{col}} \tag{1.11}$$

(u_0 is the speed of sound; $\boldsymbol{q}$ is the momentum of a phonon). Using expression (1.7') for the Hamiltonian H' of the electron-phonon interaction, after transformation to the new Fermi operators which describe the superconducting state, we obtain the following expression for the collision integral:

$$\left(\frac{\partial N}{\partial t}\right)_{\rm col} = \int |V|^2 \left\{\left(1 + \frac{\xi\xi' - \Delta^2}{\varepsilon\varepsilon'}\right) \times \right.$$
$$\times [f'(1-f)(N+1) - Nf(1-f')]\,\delta(\varepsilon' - \varepsilon - \hbar\omega) +$$
$$+ \frac{1}{2}\left(1 - \frac{\xi\xi' - \Delta^2}{\varepsilon\varepsilon'}\right)[(N+1)ff' - N(1-f)(1-f')] \times$$
$$\left. \times \delta(\varepsilon' + \varepsilon - \hbar\omega)\right\} \frac{p^2 dp\, do}{4\pi^2\hbar^3}. \qquad (1.12)$$

The collision integral (1.12), like (1.8), differs from the usual expression by the factor $(1 \pm (\xi\xi' - \Delta^2)/\varepsilon\varepsilon')$, which is related to the factors $(u_k u_{k'} - v_k v_{k'})$ and $(u_k v_{k'} + u_{k'} v_k)$ in the Hamiltonian (1.7').

We seek the distribution function in the form

$$N = N_0 - r(x)\frac{q_x}{T}\frac{\partial N_0}{\partial x}, \qquad x = \frac{\hbar\omega}{T}.$$

According to (1.2) and (1.4), the electron distribution function is

$$f = f_0 + \frac{p_x}{m}\frac{\partial f_0}{\partial \varepsilon}\frac{\varepsilon}{T}\tau_0\frac{\xi}{|\xi|}\frac{\partial T}{\partial x}, \qquad f_0 = [\exp(\varepsilon/T) + 1)]^{-1}.$$

After the substitution of the expressions for $(\partial N/\partial t)_{\rm col}$, N, and f, the kinetic equation (1.11) becomes

$$\int |V|^2 \left\{\left(1 + \frac{\xi\xi' - \Delta^2}{\varepsilon\varepsilon'}\right) \times \right.$$
$$\times \left[-(p_x + q_x)\frac{\varepsilon'}{T}\frac{\xi'}{|\xi'|}\frac{\partial T}{\partial x}\frac{\tau_0}{mT} + p_x\frac{\varepsilon}{T}\frac{\xi}{|\xi|}\frac{\partial T}{\partial x}\frac{\tau_0}{mT} - rq_x\frac{1}{T}\right] \times$$
$$\times f_0(\varepsilon) f_0(\varepsilon') e^{\frac{\varepsilon'}{T}}\delta(\varepsilon' - \varepsilon - \hbar\omega) + \frac{1}{2}\left(1 - \frac{\xi\xi' - \Delta^2}{\varepsilon\varepsilon'}\right) \times$$
$$\times \left[-(q_x - p_x)\frac{\varepsilon'}{T}\frac{\xi'}{|\xi'|}\frac{\partial T}{\partial x}\frac{\tau_0}{mT} - p_x\frac{\varepsilon}{T}\frac{\partial T}{\partial x}\frac{\tau_0}{mT}\frac{\xi}{|\xi|} - q_x r\frac{1}{T}\right] \times$$
$$\left. \times f_0(\varepsilon) f_0(\varepsilon') e^{\frac{\hbar\omega}{T}}\delta(\varepsilon' + \varepsilon - \hbar\omega)\right\} \frac{p^2 dp \sin\vartheta\, d\vartheta\, d\varphi}{4\pi^2\hbar^4} =$$
$$= N e^{\frac{\hbar\omega}{T}}\frac{\hbar\omega}{T^2} u_0 \frac{q_x}{q}\frac{\partial T}{\partial x}. \qquad (1.11')$$

We next integrate (1.11') over the angles with the aid of a relationship which follows from the law of momentum conservation:

$$\frac{d\cos\vartheta}{d\varepsilon'} = \frac{m}{pq}\frac{|\xi'|}{\varepsilon'}$$

(the polar axis is chosen in the direction of $\boldsymbol{q}$). We also note that transitions for which $\xi\xi' > 0$ make the main contribution to the scattering process, and those for which $\xi\xi' < 0$ contribute mainly to the creation of pairs of excitations (see Sec. 3). For the function $r(x)$, which determines an increment to the phonon distribution function, we now obtain the following expression (the same result is obtained by substitution of (1.6)):

$$r(x) = \frac{N_0 u_0^2}{T}\left[\frac{m^2 |V|^2}{4\pi^2\hbar^4 q}\int_{-\infty}^{\infty}\left(1 + \frac{\xi\xi' - \Delta^2}{\varepsilon\varepsilon'}\right)\times\right.$$

$$\times f_0(\varepsilon) f_0(\varepsilon + \hbar\omega) e^{\frac{\varepsilon}{T}} \frac{\varepsilon'}{|\xi'|} d\xi +$$

$$\left. + D(x)\frac{m|V|^2}{8\pi^2\hbar^4 q}\int_{-\xi_1}^{\xi_1}\left(1 - \frac{\xi\xi' - \Delta^2}{\varepsilon\varepsilon'}\right) f_0(\varepsilon) f_0(\hbar\omega - \varepsilon)\frac{\varepsilon'}{|\xi'|} d\xi\right]^{-1}\frac{\partial T}{\partial x},$$

$$D(x) = \begin{cases} 1 & \text{for } x \geqslant 2b, \\ 0 & \text{for } x < 2b; \end{cases} \quad b = \frac{\Delta}{T}, \quad x = \frac{\hbar\omega}{T},$$

$$\varepsilon' = \hbar\omega - \varepsilon, \quad \xi_1 = (\hbar\omega(\hbar\omega - 2\Delta))^{1/2}.$$

As above (see Sec. 3), we assume in the calculation of the thermal conductivity of a pure superconductor that

$$L \approx 2, \qquad L = [1 \pm (\xi\xi' - \Delta^2)/\varepsilon\varepsilon']\frac{\varepsilon\varepsilon'}{|\xi||\xi'|}. \tag{1.13}$$

This gives an error of approximately $\Delta^2 (\hbar\omega)^2/4(\varepsilon\varepsilon' \mp \Delta^2)\varepsilon\varepsilon'$ (about 0.1 for the heat flux when $\Delta/T = 1.5$, and even less for other temperatures). With the aid of this approximation, it is possible to obtain an analytic expression for the thermal conductivity. Integrating over the energy ε, we find the following expression for the function $r(x)$:

$$r \approx \frac{\text{const}}{T^2}[2x - 2\ln(e^{b+x} + 1)(e^b + 1)^{-1} +$$

$$+ D(x)(2b - x + 2\ln(e^{x-b} + 1)(e^b + 1)^{-1})]^{-1}\frac{\partial T}{\partial x}.$$

We have allowed for the fact that phonons create pairs of excitations only when $\hbar\omega \geqslant 2\Delta$. This phenomenon represents a so-called "threshold" effect. A similar effect is also observed in the study of sound attenuation in superconductors (see Chapter 3, Secs. 1, 5).

We now calculate the heat flux in the lattice:

$$Q = \sum_q N_q u_0 \frac{q_x}{|q|} u_0 q = \frac{T^4}{6\pi u_0^3}\int_0^{\infty}\frac{x^4 r(x) e^x}{(e^x - 1)^2} dx.$$

Evaluating the integral, we get the final expression for the lattice conductivity:

$$\varkappa_{pe}^s = BT^2 F(T), \tag{1.14}$$

$$F(T) \approx -8b^3(b + 1)(e^b - 1)^{-1} + 6\zeta(3)(e^b + 1) -$$

$$-6(e^b+1)\sum_s s^{-3}e^{-2bs}(2b^2s^2+2bs+1)+$$

$$+6\zeta(4)(e^b-1)-2(e^b-1)\sum_s s^{-4}e^{-2bs}(4b^3s^3+6b^2s^2+$$

$$+6bs+3)+32b^3(e^{2b}-1)^{-1}-$$

$$-a^4\sum_s\{se^{-2bs}\,\mathrm{Ei}\,[-s(2b-a)]\}+6\sum_s s^{-3}e^{-2bs}$$

$$\left(a\approx 2b-0.16\ \ (\text{for } T\to T_c\ a\to 2b),\ \ \zeta(s)=\sum_{n=1}^{\infty}n^{-s}\right),$$

$$\varkappa_{pe}^n=BT^2F(T_c).$$

The theoretical curve constructed according to (1.14) and experimental data for an In—Tl alloy are given in Figure 3.

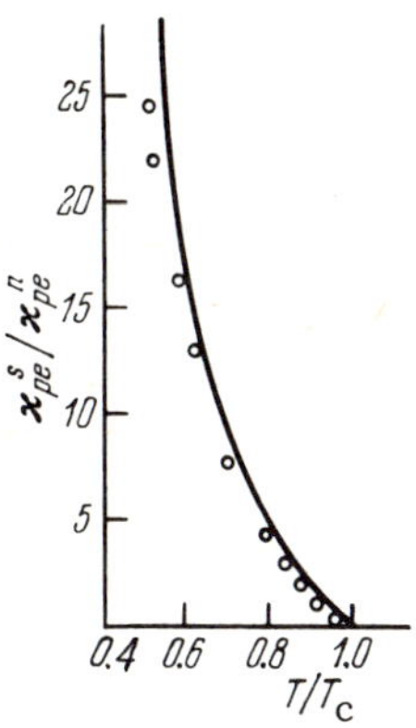

FIGURE 3. Temperature dependence of $\varkappa_{pe}^s$. Experimental data for an In—Tl alloy from /33/; the solid curve is theoretical /31/.

As in the previous cases, the ratio $\varkappa_{pe}^s/\varkappa_{pe}^n$ turns out to be a universal function of the temperature. Later, in /16/, independently of /31/, $\varkappa_{pe}^s$ was found by a numerical calculation, and the increase of $\varkappa_{pe}^s/\varkappa_{pe}^n$ with decreasing temperature was established. However, the analytic expression (1.14) (see Figure 3) more accurately describes the experimental data.

For $\Delta=0$, relationship (1.14) becomes the familiar equation from the quantum theory of metals (see, for example, /13/) which describes the contribution of the lattice thermal conductivity $\varkappa_{pe}^n$, which decreases with decreasing temperature. This decrease, as we know, is associated with a corresponding decrease in the number of thermal phonons.

In superconductors, as opposed to normal metals, $\varkappa_{pe}^s$ increases with decreasing temperature, as is seen in (1.14). This peculiarity is associated with the increase of the phonon free path because of the exponential decrease in the number of electron excitations (the phonon free path is determined by the scattering of phonons by electron excitations, $l_{pe} \sim (n_e\sigma)^{-1}$, where n_e is the number of electron excitations and σ is the scattering cross section).

At a sufficiently low temperature, the phonons begin to be scattered mainly by lattice defects and crystal boundaries. Here the temperature dependence of the thermal conductivity is determined by the laws governing normal solids at low temperatures.

For $T \ll \Theta$ (Θ is the Debye temperature), the phonon-phonon interaction has no effect. Introduction of Umklapp processes, which are significant in the study of anharmonic terms at low temperatures (for further details see, for example, Peierls's monograph /34/), results in an expontially decreasing contribution of the phonon-phonon interaction to the heat flux. The scattering of phonons by lattice defects and crystal boundaries, which plays a major role for $T \to 0$, makes the thermal conductivity $\varkappa_p \approx \varkappa_{pd}$ go to zero as $T \to 0$. A detailed theory of lattice conductivity in normal solids is presented, for example, in Ziman's monograph /35, Chapter 8/. When the phonon free path is constant (such as when the phonons are scattered by the specimen boundary), $\varkappa_{pd}$ is proportional to T^3.

Thus, as $T \to 0$, the lattice conductivity decreases and generally approaches zero as a power function. We have seen that the contribution from the scattering of phonons by electrons in superconductors is described by a function which increases when the temperature falls below T_c. Consequently, in the intermediate temperature range, when $\varkappa_{pe}^s \approx \varkappa_{pd}$, a maximum lattice conductivity should be observed. The maximum is indeed observed experimentally (see below).

5. COMPARISON WITH EXPERIMENT

The various forms of temperature dependence (1.5), (1.10), and (1.14), which have been developed above, enable us to explain to a great extent the currently available experimental data on thermal conductivity of superconductors /32/.

As noted in Sec. 1, there are six mechanisms of heat conduction, and the relative contribution of each mechanism is highly sensitive to the experimental conditions.

For pure superconductors, as well as for specimens containing a low impurity concentration, $\varkappa^s$ is approximately equal to $\varkappa_e^s$ (for very pure superconductors $\varkappa^s \approx \varkappa_{ep}^s$, and for weakly doped superconductors $\varkappa^s \approx \varkappa_{ed}^s$), with the exception of the temperature range close to the absolute zero. The thermal conductivity falls with decreasing temperature according to (1.5) and (1.10) (Figure 4, curve 1). This fall is initially slow, but later becomes exponential due to the exponential decrease in the number of electron excitations.*

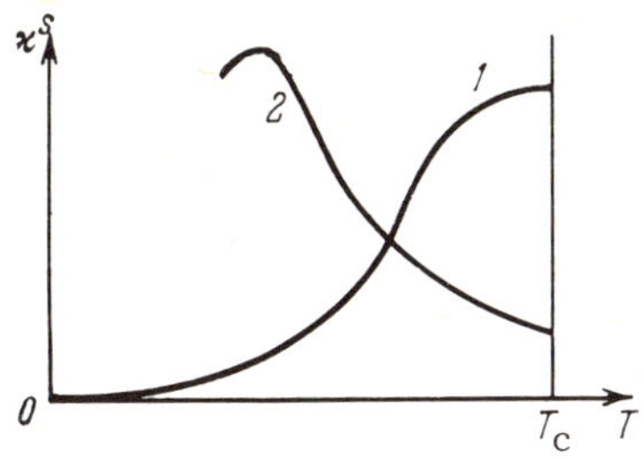

FIGURE 4. Temperature dependence of the thermal conductivity in superconductors: 1) electronic conductivity; 2) lattice conductivity.

For a qualitative evaluation we will use a simple formula from the kinetic theory, $\varkappa = c_V lv/3$ (c_V is the specific heat, l is the free path length, and $v = v_F$). In the case which interests us, $c_V \sim n \sim \sim \exp(-\Delta/T)$, but l varies according to a power function.

The above dependence describes sufficiently well the experimental data on thermal conductivity of Al and Zn /36, 38/, Tl /15/, Sn /15, 23/, Va /39/, In–Hg /40/, and In /24/ (for a review of the experimental data see also /5, 8, 11/).

As can be seen in (1.5) and (1.10), the electronic conductivity $\varkappa_{ed}^s$, which is due to the scattering of electrons by impurities, falls more slowly than $\varkappa_{ep}^s$ as the temperature decreases below T_c (compare Figures 1 and 2). Thus, an increase in the concentration of the impurity must result in a somewhat slower decrease of the function $\varkappa^s(T)$. The relative effect of the impurity is determined by the ratio of the parameters α and β in expression (1.1) for the thermal conductivity of a normal metal.

Near the critical temperature, the increase in the effect of impurities on the electronic conductivity is primarily determined by the relationship $a = \alpha T_c^3/\beta$, which is found directly from the experimental data on the thermal conductivity of normal metals. Thus, for example, in a study of In–Hg alloys /40/, an impurity concentration of $n = 0.1$ at.% Hg corresponds to $a = 0.157$; for $n = 0.2$ at.% Hg, $a = 0.100$; for $n = 0.5$ at.% Hg, $a = 0.046$.

Guenault /41/ studied experimentally the rate of decrease of the electronic thermal conductivity as a function of the impurity concentration. Three tin specimens of different degrees of purity were employed for the measurements. It turned out that in the superconducting state, the smallest decrease of the thermal conductivity

* Note that in the temperature range close to the critical temperature, electron-electron scattering $\varkappa_{ee}^s$ may have some effect.

was observed for the Sn specimen containing the highest impurity concentration (the largest coefficient β). This result is consistent with the previous conclusion concerning the slower drop in $\varkappa_{ed}^{s}$ in comparison with $\varkappa_{ep}^{s}$ (compare Figures 1 and 2).

Thus, in pure superconductors and in superconductors with a low impurity concentration, the electronic thermal conductivity (excluding the temperature range $T \to 0$) plays the major role, as it does in normal metals. A decrease in the number of electron excitations results in an abrupt drop in thermal conductivity when the temperature decreases below T_c. A curve describing the temperature dependence of the thermal conductivity of a metal passing into a superconducting state is characterized by lower values of the thermal conductivity than the curve for a normal metal. There may be two reasons for this downward deviation.

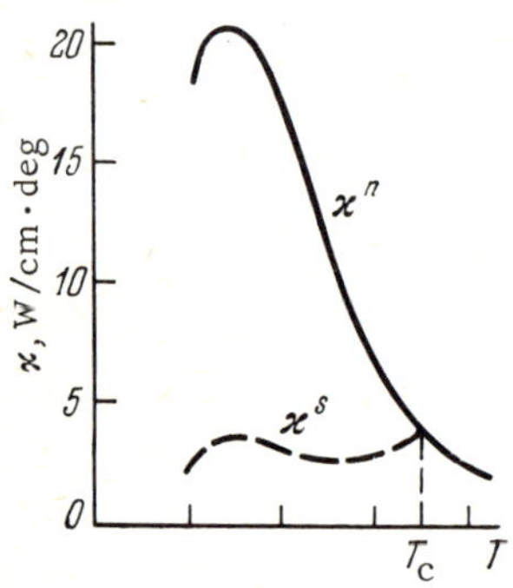

FIGURE 5. Thermal conductivity of lead in the superconducting and normal states.

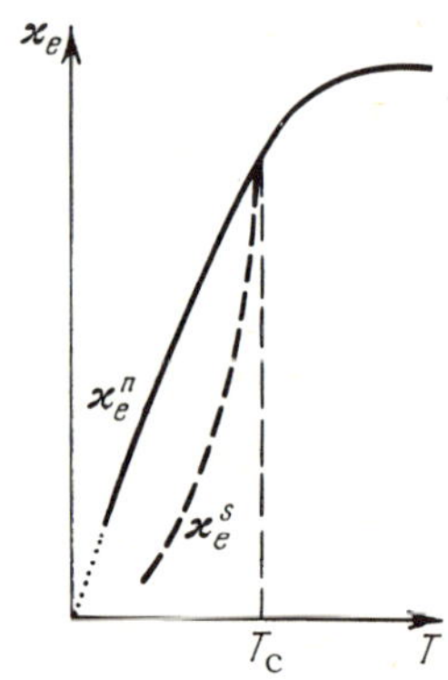

FIGURE 6. Thermal conductivity of pure tin in the superconducting and normal states.

In Sec. 1 we spoke about the existence of a maximum of thermal conductivity in normal metals, which is well known in the conventional quantum theory of metals. If the critical temperature for the transition into the superconducting state, T_c, is higher than T_{max}, the temperature at which the maximum occurs (this situation is commonly observed in sufficiently pure specimens), the downward deviation of $\varkappa_e(T)$ upon transition through T_c is quite marked (Figure 5). In specimens where T_c is close to T_{max}, the deviation in $\varkappa_e(T)$ is not as pronounced as in the first case, being more gradual (Figure 6). The impurity concentration in such specimens is higher than in the first case. However, it should be taken into account that we do not consider excessively high concentrations, since otherwise the lattice conductivity $\varkappa_p$ begins to play a major role.

At very low temperatures, $\varkappa_e^s \ll \varkappa_p^s$; then $\varkappa^s \approx \varkappa_{pd}$. The thermal conductivity is primarily determined by the scattering of phonons by impurities and defects, and their reflection from the crystal boundaries. It approaches zero when $T \to 0$, usually as a power function. For example, the reflection of phonons from crystal boundaries, which begins to play a major role at a sufficiently low temperature, results in the proportionality of $\varkappa_{pd}$ to T^3. The sudden increase in the role of the lattice conductivity is an important anomaly of the superconducting state (in normal metals $\varkappa^n$ is approximately equal to $\varkappa_e^n$ at all temperatures) and is associated with a decrease in the normal component of electron fluid. The condensate of Cooper pairs possesses zero entropy and therefore does not contribute to the heat flux.

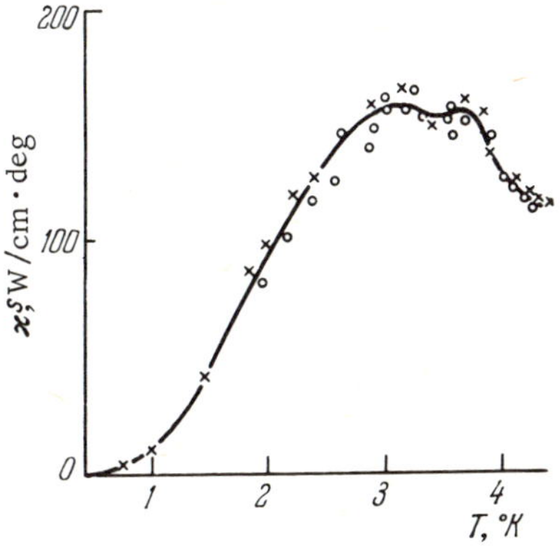

FIGURE 7. Thermal conductivity of different tin specimens /15/.

In /42/ the thermal conductivities of Pb, Sn, In, and Ta were studied in the low-temperature range; in /43/ the properties of Sn were studied; in /44/ those of Sn and In, and in /15, 45–47, 61/ those of Tl, Sn, Al, Ga, Zn, Cd, and Pb. It was actually established that for $T \to 0$ (e.g., for Tl, which was studied by Zavaritskii /15/ at $T \lesssim (0.3–0.4)$°K), a superconductor does not differ from a dielectric in its heat transfer properties. It can be seen in Figure 7 that as T approaches zero, the nature of the temperature dependence of $\varkappa^s$ in Sn changes /15/; the thermal conductivity begins to obey the law $\varkappa_{pd} \sim T^3$.

The fact that the phonon thermal conductivity begins to play the major role as the temperature is decreased is also experimentally confirmed in the following way. In plastic deformation of a specimen the free path of the long-wave thermal phonons is altered, as they begin to be scattered mainly by the dislocations which arise in such a deformation. For the electrons, which have shorter wavelengths, the appearance of dislocations does not have a substantial effect, and they continue to be scattered primarily by the impurity atoms. Therefore, the thermal conductivity of a specimen should decrease during plastic deformation if the phonons make the main contribution to the heat flux, and practically does not change if the electronic component plays the major role.

Measurements performed by Mendelssohn and Montgomery /48/ showed a sixfold decrease in the thermal conductivity of superconducting lead following plastic deformation at $T = 1°K$. This also confirmed our understanding of the increased role of the phonon thermal conductivity associated with the scattering of phonons by crystal defects, impurities, and crystal boundaries, as the temperature approaches zero.

Heavy doping of a specimen decreases both the free path of the electrons and the electronic component of the conductivity according to the equation $\varkappa = c_V l v/3$; $\varkappa_p^s$ is much greater than $\varkappa_e^s$ at all temperatures. In the intermediate temperature range ($T \approx (0.3 - 0.5)\,T_c$) and as T approaches T_c the scattering of phonons by electrons is the dominant process, and the conductivity is described by equation (1.14) (Figure 4, curve 2). The value of $\varkappa_{pe}^s$ rises with decreasing temperature. This explains the experimentally observed anomalous behavior of the thermal conductivity of superconducting alloys. In Figure 3 experimental data for the In—Tl alloy are given /33/. An increase in $\varkappa_p^s$ in Nb was observed at $T \approx 0.4\ T_c$ /49, 50/. An increase in $\varkappa_p^s$ with decreasing temperature is also mentioned in /51, 52/.

The lattice conductivity of NbC is studied in /53/. As the authors note, the experimental results are compatible with the function obtained in /16/ if the value of the ratio $2\Delta(0)/T_c$ is set equal to 4 instead of 3.5 as the BCS theory would require. Such a value for $2\Delta(0)/T_c$ is possible in the strong coupling theory (see Sec. 9). In this context we note that the function $\varkappa(T)$ at temperatures close to T_c is not determined directly by the ratio $\Delta(0)/T_c$, but by the coefficient α in the equation

$$\Delta(T)_{T \to T_c} = \alpha(1 - T/T_c)^{1/2}.$$

Meanwhile, as indicated in /53/, the ratio $T_c/\tilde{\omega} \lesssim 0.01$ for NbC is too small to consider it a strong coupling superconductor.

The contradiction is apparently not due to the presence of strong coupling in this superconductor. It is more likely due to the fact that the theoretical curve obtained in /16/ by numerical methods qualitatively describes the increase in $\varkappa_{pe}^s$ as the temperature decreases below T_c, but the increase in /16/ is too slow to fit equation (1.14) (see Figure 3). In experiment /53/ (see also /40/), however, one observes a sharper dependence of $\varkappa_{pe}^{s}$ on T than in /16/. This result fits equation (1.14), which was obtained by an analytic solution of the kinetic equation.

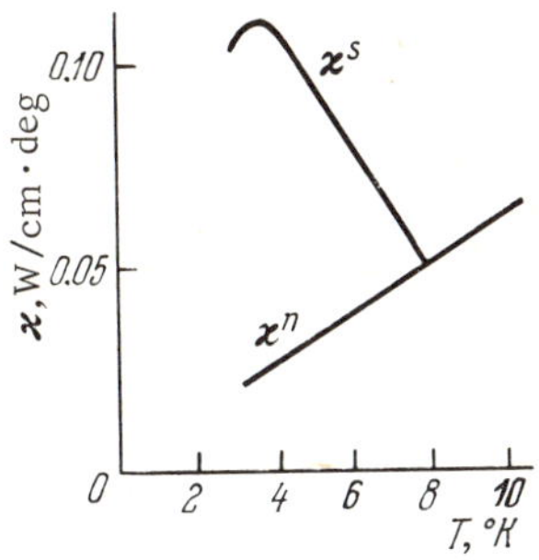

FIGURE 8. Thermal conductivity of the Pb+10 at.% Bi alloy /45/.

Figure 8 gives the temperature dependence of the thermal conductivity of the Pb + 10 at. % Bi alloy. At first the function $\varkappa^s(T)$ rises with decreasing temperature according to (1.14), and after passing through a maximum it begins to decrease as T goes to zero. A maximum lattice conductivity is also observed in /47, 54/. In the temperature range corresponding to the maximum, $\varkappa_{pe}^{s}$ is approximately equal to $\varkappa_{pd}$ (see Sec. 4). During a further drop in temperature, the phonons are primarily scattered by impurities, which produces a decrease in $\varkappa^s(T)$.

In the intermediate case of not very pure superconductors, $\varkappa_e^s$ plays the major role near T_c and thus $\varkappa^s$ falls with decreasing temperature (see curve 1 in Figure 4) according to equation (1.5). At sufficiently low temperatures $\varkappa_p^s$ becomes larger than $\varkappa_e^s$ and the subsequent behavior of the conductivity is described by curve 2 in Figure 4, which corresponds to equation (1.14). Such a dependence was observed experimentally in Sn /43/, Hg /55/, Pb /56/, and In–Pb alloys /37/. Figure 9 shows how a gradual increase in the concentration of an impurity results in an increase in the contribution of the lattice conductivity and the appearance of a maximum.

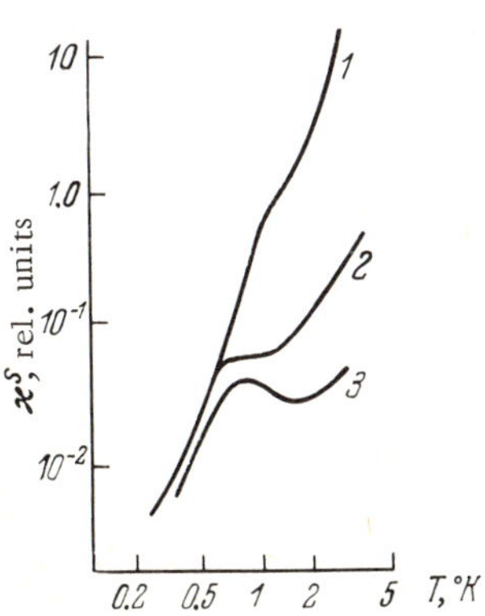

FIGURE 9. Changes in the function $\varkappa^s(T)$ with increasing impurity concentration:

1) pure Sn; 2) Sn + 0.3 at.% In; 3) Sn + 3 at.% In.

The dependence of the conductivity on an applied external magnetic field, as well as on the temperature gradient, was studied in /57—59/. Naturally, thin films of thickness less than the penetration depth were the subject of the study (in /58/ films of indium with a thickness $L \approx 650$ Å and tin with $L \approx 700$ Å were studied).

As we know, an increase in the magnetic field reduces the energy gap. Therefore, an increase in the magnetic field is equivalent to an increase in temperature. We should thus expect an increase in the electronic thermal conductivity $\varkappa_e^s$ and conversely, a decrease in the lattice component $\varkappa_p^s$ as the external magnetic field rises. In fact, this was observed in the previously cited work /57, 59/. In the temperature range near the critical point, where $\varkappa_{ed}^s$, caused by the scattering of electrons by impurities and the boundaries, is the major factor, an increase in the function $\varkappa^s(H)$ in the thin films was observed with an increase in H. In the temperature range where the phonon thermal conductivity $\varkappa_{pe}^s$ is the major factor, a decrease in the conductivity $\varkappa^s(H)$ was actually observed as H was increased.

In Figure 5 the temperature dependence of the conductivity of superconducting doped lead obtained in /56/ is given. The fall in $\varkappa^s(T)$ just below T_c is associated with a corresponding decrease in the value of $\varkappa_e^s$ which is the major factor near T_c. As can be seen in the figure, the conductivity of Pb then begins to rise, and finally a maximum is observed, associated with a maximum of $\varkappa_p^s$.

6. BASIC TYPES OF CONDUCTIVITY IN SUPERCONDUCTORS

The theoretical and experimental studies of the processes of heat transfer in superconductors lead to the conclusion that there are three types of temperature dependence for their thermal conductivity.

Pure and slightly doped superconductors show the first type (Figure 10, a). In these the electronic thermal conductivity plays the major role as T approaches T_c and in the intermediate temperature range. With a decrease in temperature, $\varkappa^s$ declines on the whole exponentially according to equations (1.5) and (1.10). The former equation corresponds to slightly doped superconductors, and the latter to pure ones. In Figure 1 the temperature dependence of $\varkappa^s$ for thallium (first type of conductivity) is shown. As $T \to 0$ the lattice thermal conductivity $\varkappa_{pd}$ begins to be the major factor, and $\varkappa^s$ starts approaching zero as a power function.

Superconductors with a high impurity concentration show the second type (Figure 10, b). The thermal conductivity of the strongly doped superconductors, as opposed to substances of the first type, increases just below T_c as the temperature is decreased. In this case the lattice thermal conductivity $\varkappa^s_{pe}$ described by (1.14) is the major factor. The characteristic features of this case is the presence of a maximum conductivity and after passing through the maximum $\varkappa^s(T)$ approaches zero with the function $\varkappa_{pd}(T)$. The Pb + 10 at.% Bi alloy (Figure 8) is an example of a superconductor with conductivity of the second type.

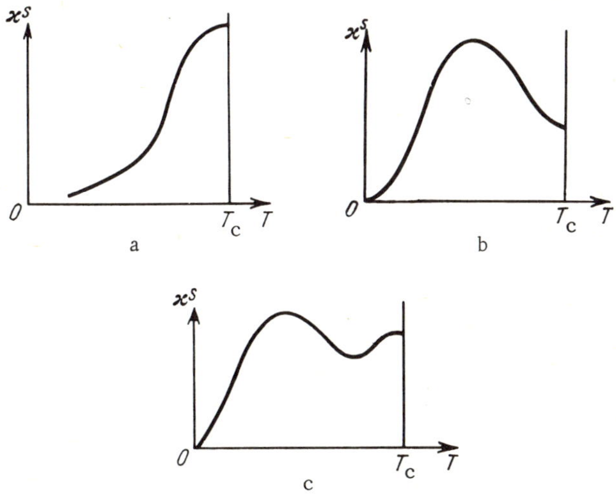

FIGURE 10. The basic types of thermal conductivity:

a) pure superconductors; b) superconductors with a high impurity concentration; c) the intermediate case.

Finally, the third type of thermal conductivity (Figure 10, c), which describes the behavior of moderately doped superconductors, is a combination of the first type (when $T \approx T_c$) and the second (in the intermediate temperature range). The data given in /15, 16, 56/ for Pb and Sn (Figure 5) suggests that the corresponding specimens are examples of superconductors related to the intermediate case. In these $\varkappa^s_{ed}$ is the major factor near T_c, and therefore the conductivity initially falls with decreasing temperature according to equation (1.5). However, $\varkappa^s_p$ next becomes greater than $\varkappa^s_e$, and near absolute zero these superconductors behave like substances with conductivity of the second type.

7. EFFECT OF ANISOTROPY ON THE THERMAL CONDUCTIVITY OF SUPERCONDUCTORS

In our study of the various mechanisms of heat conduction so far, an isotropic model was employed. A square-law electron dispersion was assumed. The electron-phonon interaction was described by a constant value, independent of the direction.

Real metals are characterized by a substantially anisotropic Fermi surface. The energy gap $\Delta(\boldsymbol{k})$ is generally a function of the orientation. However, the anisotropy of the energy gap is not large, and the isotropic BCS model, as we know, describes the experimental data fairly well.

Heat conduction is an integral effect. Therefore, the averaging used in the calculations does not generally lead to a substantial anisotropy of the thermal conductivity. The equations given in Secs. 2–4 are obtained from an isotropic model, and therefore they fit the experimental data fairly well. Uniaxial crystals (such as Ga and Zn) are an exception. In these superconductors, Zavaritskii /46, 61/ observed a marked difference in the form of the function $\varkappa(T)$ for different crystal orientations.

The question of the electronic conductivity of uniaxial superconductors is discussed in a paper by Khalatnikov /60/. In the anisotropic case, the kinetic equation which describes the behavior of electron excitations interacting with impurities has the form

$$\frac{f}{T^2}\xi_{\boldsymbol{k}}\left(\frac{\partial\xi_{\boldsymbol{k}}}{\partial\boldsymbol{k}}\nabla T\right)=I_{\text{col}} \tag{1.15}$$

(I_{col} is the collision integral; $\varepsilon_{\boldsymbol{k}}=(\xi_{\boldsymbol{k}}^2+\Delta^2(\boldsymbol{k}))^{1/2}$.

Near the critical temperature, the energy gap is small, and the effects associated with the anisotropy are not significant. The situation is different at sufficiently low temperatures which satisfy the condition

$$T\ll\Delta_{\text{min}}.$$

In this temperature range the energy gap is virtually independent of T, and the distribution function of the electron excitations is exponential, $\exp(-\Delta/T)$.

In the anisotropic case, the energy gap $\Delta(\boldsymbol{k})$ is minimum in certain extremum directions. In a uniaxial crystal with an extremum direction which coincides with its principal axis of symmetry, the majority of the electron excitations will move in the direction of this axis.

The collision integral appearing in equation (1.15) can be written in the form (see Sec. 2)

$$I_{\text{col}} = -2\pi \int |V|^2 (u_k u_{k'} - v_k v_{k'})^2 \delta(\varepsilon_k - \varepsilon_{k'})(f_k - f_{k'})\, d\tau_k,$$
$$u_k^2 = {}^1/_2 (1 + \xi_k/\varepsilon_k), \quad v_k^2 = {}^1/_2 (1 - \xi_k/\varepsilon_k). \tag{1.16}$$

In the anisotropic case, only the energy of the electron excitations is conserved in elastic scattering by impurities. The absolute value of the momentum is not conserved in distribution from the isotropic case.

We seek the distribution function f_k in the form

$$f_k = f_k^0 (1 + \chi_k).$$

After integration over ξ_k the kinetic equation becomes

$$-\frac{\varepsilon_k}{T^2}\left(\frac{\partial \xi_k}{\partial k}\nabla T\right) = \frac{\pi}{2}\int |V|^2 \frac{\xi_k + \xi_{k'}}{\xi_k |\xi_{k'}|}\left(\frac{\partial^2 \tau}{\partial \xi_k\, do'}\right)(\chi_k - \chi_{k'})\, do'. \tag{1.17}$$

Equation (1.17) is solved for the extremum directions. We denote the thermal conductivity along the extremum direction and the direction perpendicular to it by $\varkappa_{\parallel}^s$ and $\varkappa_{\perp}^s$. We introduce the quantities V_0 and V_π which are the matrix elements for scattering at the angle 0 and π, respectively, for excitations initially moving in the extremum direction.

Solving equation (1.17) leads to the following expressions for $\varkappa_{\parallel}^s$ and $\varkappa_{\perp}^s$:

$$\varkappa_{\parallel}^s \approx \frac{8\Delta^2}{5\pi T}\left(\frac{\partial \xi_k}{\partial k}\right)_F \frac{1}{1{,}8V_\pi^2 + 0{,}2V_0^2}\exp\left(-\frac{\Delta}{T}\right), \tag{1.18}$$

$$\varkappa_{\perp}^s \approx \frac{10\Delta^2}{3\pi\Delta''}\left(\frac{\partial \xi_k}{\partial k}\right)_F \frac{1}{V_0^2 + V_\pi^2}\exp\left(-\frac{\Delta}{T}\right) \tag{1.18'}$$

(Δ'' is the second derivative of the energy gap with respect to the angle θ between ∇T and the vector $\boldsymbol{k}$ near the extremum). The exponential part of the temperature function for $\varkappa^s$, according to (1.18) and (1.18') is the same in every direction. The ratio $\varkappa_{\perp}^s/\varkappa_{\parallel}^s$ is equal to

$$\frac{\varkappa_{\perp}^s}{\varkappa_{\parallel}^s} \approx \frac{2T}{\Delta''}\cdot\frac{1{,}8V_\pi^2 + 0{,}2V_0^2}{V_\pi^2 + V_0^2},$$

i. e., the conductivity in the direction perpendicular to the extremum direction is approximately T/Δ times smaller than the extremum value.

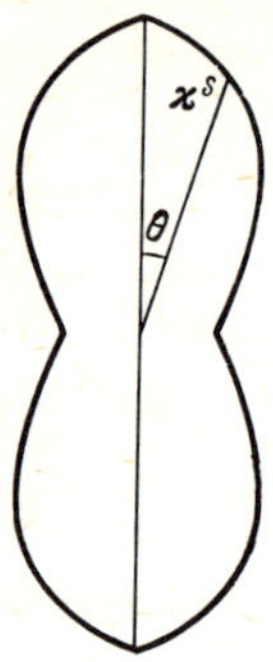

FIGURE 11. Thermal conductivity as a function of the angle θ /60/.

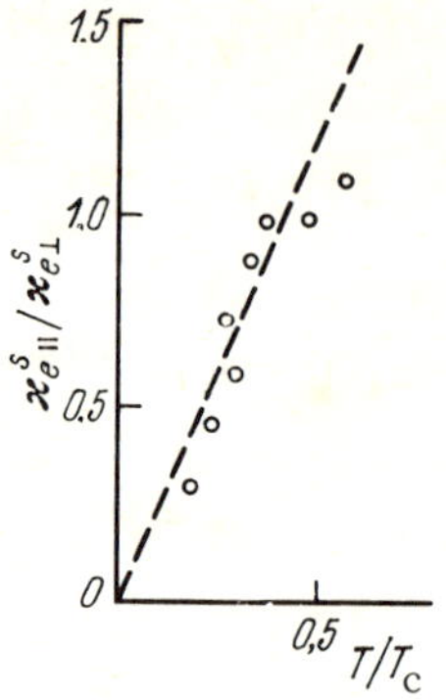

FIGURE 12. Dependence of $\varkappa^s_{e\parallel}/\varkappa^s_{e\perp}$ on the temperature /46/. $\varkappa^s_{e\parallel}$ and $\varkappa^s_{e\perp}$ are the thermal conductivities along and perpendicular to the hexagonal axis. The broken line is theoretical /60/.

In Figure 11 the thermal conductivity as a function of the angle θ (when $\varkappa^s_{\perp}/\varkappa^s_{\parallel} = {}^1/_5$) is depicted. The angle is measured from the extremum direction, and the function has the form

$$\varkappa^s = \varkappa^s_{\parallel}\cos^2\theta + \varkappa^s_{\perp}\sin^2\theta.$$

Anisotropy in the conductivity has been experimentally detected in crystals of Ga /61/, Zn, and Cd /46/. In /46/ a detailed comparison of Zn with theory /60/ is given. As can be seen in Figure 12, the agreement is quite good. The anisotropy of the energy gap in Zn is characterized by the following data:

$$\Delta_{\min} = 1.2\ T_c;\ \Delta_{\max} - \Delta_{\min} \approx 0.55\ T_c.$$

We will discuss the influence of the energy gap anisotropy on the properties of superconductors in greater detail in Chapter III, which is devoted to sound attenuation in superconductors. At this point we note that the effects due to anisotropy or to overlapping bands are significant only in fairly pure specimens. The free path must satisfy Anderson's condition /30/ that $l \gg \xi_0$.

In this context we should mention a paper /62/ in which the temperature dependence of the electronic thermal conductivity $\varkappa^s_{ed}$ associated with the scattering of electrons by impurities and crystal defects was experimentally studied in superconducting Al. At impurity concentrations corresponding to free paths of $l_1 = 4\cdot10^{-5}$ cm and $l_2 = 6\cdot10^{-4}$ cm, the function $\varkappa^s_e(T)$ is described fairly well by equation (1.5). However, as the author of /62/ pointed out, for a specimen with $l_3 = 5.6\cdot10^{-3}$, the experimental data consistently fell above the experimental curve for $\varkappa^s/\varkappa^n$. Since Anderson's condition $l \gg \xi_0$ was fulfilled in this case, the retardation in the fall of the conductivity mentioned in /62/ can apparently be explained by the anisotropy of the energy gap or by a multiband structure. Clearly, when the decrease is exponential, the minimum energy gap Δ_{min} begins to be a major factor. The electronic thermal conductivity decreases more slowly with decreasing temperature than in the isotropic, single-band model. Therefore, the experimental points obtained in /62/ for the least doped specimen lie above the theoretical curve.

A preferred direction can be artificially created in a superconductor by placing it in an external magnetic field. The influence of a magnetic field on the thermal conductivity of superconductors was theoretically discussed in a paper by Maki /17/. Naturally, this study dealt with films whose thickness was smaller than the penetration depth. It was shown that the introduction of the magnetic field results in anisotropy of the thermal conductivity.

Anisotropy of the thermal conductivity in a magnetic field was observed experimentally in /63, 64/. In a weak magnetic field, $H < H_{c1}$, it was found that $\varkappa_{\parallel}^{s} < \varkappa_{\perp}^{s}$ ($\varkappa_{\parallel}^{s}$ is the thermal conductivity when the field and the heat flux are parallel to each other, and $\varkappa_{\perp}^{s}$ corresponds to the case in which the field and heat flux are mutually perpendicular).

In /63/ the thermal conductivity of In + 2 at. % Bi alloys was studied in a magnetic field. It was noted that in a weak magnetic field $\varkappa_{\perp}^{s} - \varkappa_{\parallel}^{s} = AH^2$, where the value of A was independent of the parameters of the specimen. In a strong magnetic field, when a mixed structure arises, the sign of the effect was reversed. In this case, $\varkappa_{\parallel}^{s} > \varkappa_{\perp}^{s}$.

8. THERMAL CONDUCTIVITY OF SUPERCONDUCTORS WITH A STRONG ELECTRON-PHONON INTERACTION

An experimental investigation of the thermal conductivity of pure superconducting specimens of Pb and Hg /55, 81/ showed that $\varkappa^s(T)$ falls with decreasing temperature near T_c more abruptly than in other metals (Figure 13). The conductivity of these superconductors is inadequately described by equations (1.10) and (1.10'). This peculiarity of lead and mercury is associated with their belonging to the group of so-called anomalous superconductors. Nb, Ga, In, NbN, films of certain metals (such as Bi films), and several alloys (Nb_3Sn, Nb–Mo, $LaSN_3$, etc.) belong to the same group. In these substances the electron-phonon interaction which determines the superconductivity is not weak; and therefore their properties (see, for example, the review /5/) are poorly described by the usual theory developed in the weak coupling approximation. The anomalous superconductors are characterized by a significantly larger ratio T_c/Θ than other metals. For example, for Pb, $T_c/\Theta \approx 0.1$ (for comparison we mention that for Al, $T_c/\Theta \approx 1/400$); according to the formula $T_c \approx \Theta \exp(-1/g)$, this yields for the coupling constant $g_{Pb} \approx 0.4$.

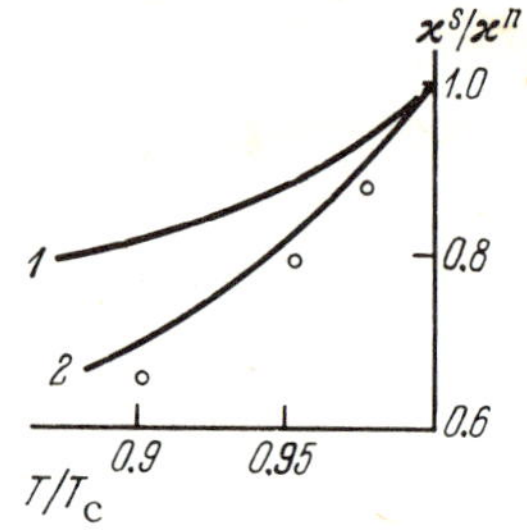

FIGURE 13. Thermal conductivity of weak coupling superconductors. The experimental points were obtained for Pb. Curve 1 corresponds to the BCS theory, and curve 2 is based on the theory of strong coupling superconductors.

From equations (1.5), (1.10), and (1.14) it can be seen that the thermal conductivity of superconductors is determined by a universal function of the parameter $b = \Delta/T$. According to the usual theory /1, 4/ $(\Delta/T)_{T\to T_c} = 3.06\ (1 - T/T_c)^{1/2}$. For anomalous superconductors this relationship, which was obtained in the weak coupling approximation, is incorrect, and the experimentally observed difference between theory and experiment is associated with this fact.

A theoretical treatment of the properties of weak coupling superconductors is given in /82–84/. In /88/ the specific features of the thermal conductivity of superconductors are noted. The temperature dependence of the energy gap near T_c, which is essential for the solution of the thermal conductivity problem, was found in /84, 85/.

We will discuss below the basic premises of the strong coupling superconductor theory and show how the value of $\Delta(T)$, which determines the temperature dependence of the kinetic coefficients of interest, is calculated /82–85/.

8.1. The Energy Gap

The study of the properties of strong coupling superconductors must be based on the Fröhlich model, which directly considers the electron-phonon interaction. The initial equation for determining the energy gap is Eliashberg's equation /89/ (see also /90/) for the self-energy parts $\Sigma(\omega_n, T)$ which describes the Cooper pairs. The equation for this function can be written with the aid of diagram methods for $T \neq 0$ and has the form

$$\Sigma(\omega_n, T, \boldsymbol{p}) = \frac{T}{(2\pi)^3} \sum_{\omega_{n'}} \int d\boldsymbol{k} g^2 \frac{\omega^2(\boldsymbol{q})}{\omega^2(\boldsymbol{q}) + (\omega_n - \omega_{n'})^2} \frac{\Sigma(\omega_{n'}; T, \boldsymbol{k})}{K}. \quad (1.19)$$

Here $\omega_n = (2n+1)\pi T$, $\omega(\boldsymbol{q})$ is the phonon energy, $\boldsymbol{q} = \boldsymbol{p} - \boldsymbol{k}$,

$$K = \omega_{n'}^2 + i\omega_{n'}(\Sigma_1^+ - \Sigma_1^-) + \xi^2 + \xi(\Sigma_1^+ + \Sigma_1^-) + \Sigma_1^+\Sigma_1^- + \Sigma^2,$$

$$\Sigma_1^{\pm} = \Sigma_1(\pm i\omega_n, T); \ \hbar = 1.$$

We do not pause here to write out the equation for the self-energy part Σ_1, which describes the scattering process.

It can be seen from (1.19) that $\Sigma(\omega_n, T)_{\omega_n \to \infty} \sim \omega_n^{-2}$. This feature of the behavior of $\Sigma(\omega_n, T)$ must be considered in choosing the zeroth approximation for the function obtained by solving (1.19). As opposed to the usual case, we do not assume that the "delay" described by the term $(\omega_n - \omega_{n'})^2$ in the denominator of the phonon Green's function is small.

We seek a solution of equation (1.19) in the form

$$\Sigma(\omega_n, T) = \Sigma_0 + \Sigma', \quad (1.20)$$

where $\Sigma' \ll \Sigma_0$, and the zeroth approximation is

$$\Sigma_0 = C(T) \frac{\omega^2}{\omega^2 + \omega_n^2} \quad (1.20')$$

(ω is the effective phonon frequency; see below).

We first consider the temperature range close to the critical point /84, 85/. Eliminating g^2 from (1.19) and introducing the critical temperature as the parameter, we obtain a relationship between $\Sigma_{T \to T_c}$ and T_c. The function $C(T) \equiv \Sigma(0, T)$ is next calculated via the approximation in (1.20) and (1.20'); when this is done, the correction $C'(T)$ due to the term Σ' is indeed small. To calculate the energy gap, analytic continuation of the thermodynamic Green's function to the upper half-plane is required.

The function $\omega(\boldsymbol{q})$, which describes the phonon spectrum of the superconductor, enters the integrand which determines both $C(T)$ and the energy gap. Therefore, in the case of strong coupling, the factor $\alpha = \Delta/T(1 - T/T_c)^{1/2}_{T \to T_c}$ is not a universal constant as it is in the weak-coupling approximation (in the BCS theory, $\alpha = 3.06$). Both its value and the temperature dependence of the energy gap are substantially dependent on the nature of the phonon spectrum.

The phonon spectra of many metals were studied either by neutron scattering (detailed measurements of the phonon spectrum of Pb were carried out by this method in /91/), or by tunneling measurements (see, for example, /92–94/). These measurements showed that the phonon density of states is usually characterized by the presence of two sharp peaks, corresponding to the transverse and longitudinal vibration branches. Therefore, the model in which the phonons are described by two Einstein-type branches with constant frequencies corresponding to the two peaks provides a sufficiently good approximation.

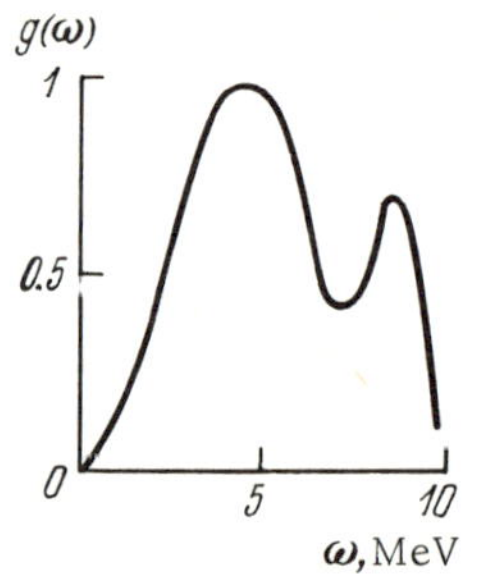

FIGURE 14. The function $g(\omega)$ for Pb /92/. $g(\omega) = \alpha^2(\omega)F(\omega)$; $\alpha^2(\omega)$ describes the electron-phonon interaction; $F(\omega)$ is the density of states.

Note that the approximation in which the phonon spectrum is described by one characteristic frequency is completely satisfactory for Pb since, first, the peaks of the phonon density of states for Pb are close to each other (Figure 14), and second, the lower peak, which corresponds to the larger correction value, is the major factor (see (1.24)). These considerations are confirmed by the fact that the analytic expression connecting $\Delta(0)$ and T_c (see below) obtained in /82, 83/ coincides with the results in /86, 87/, where numerical calculations were carried out using the exact form of the spectrum from /91/.

Knowledge of the function $C(T)$ and the identity $\omega = \Sigma(-i\omega)$ allow us to determine the energy gap,

$$(\Delta/T)_{T\to T_c} = \alpha(1 - T/T_c)^{1/2}, \qquad (1.21)$$

where $\alpha_{Pb} \approx 4$. This value is quite different from $\alpha_{BCS} = 3.06$. This difference explains many of the peculiarities of anomalous superconductors. Thus, for example, the well-known question concerning the specific-heat discontinuity during the transition from the superconducting to the normal state can be answered. The usual superconductivity theory, as is known /1, 4/, leads to the following value for the discontinuity in the isotropic case:

$$\beta = (c^s/c^n)_{T=T_c} = 2.43.$$

It was shown /95/ that consideration of the anisotropy leads to a lower discontinuity than in the isotropic case. The inequality $\beta < 2.43$ is only strengthened for the multiband superconductor model /96, 160/, in which β can be calculated by using a method developed in /95/. However, as we know, the relationship $\beta < 2.43$ contradicts the experimental data (see, for example, the review /98/ obtained in a study of several superconductors. Thus, for example, $\beta_{Pb} = 3.65$ /5/ (according to data from /11/, $\beta_{Pb} = 3.4$) $\beta_{Hg} = 3.2$, $\beta_{Nb} = 3.07$, $\beta_{Sn} = 2.6$ (but $\beta_{Zn} = 2.25$).

Results indicating that the specific heat jump cannot exceed 2.43 were obtained in the weak-coupling approximation. We will show that abandoning this approximation leads to values of β greater than 2.43, which agree with the experimental data.

The usual combinatorial expression for the entropy of a superconductor (it is sufficiently accurate for superconductors with strong coupling as well) in the case of the Fröhlich model is brought to the form

$$S^s = \frac{\pi^2}{3}\nu T\left[1 - \frac{3}{2\pi^2}\left(\frac{\Delta}{T}\right)^2\right] = \frac{\pi^2}{3}\nu T\left[1 - \frac{3}{2\pi^2}\alpha^2\left(1 - \frac{T}{T_c}\right)\right] \quad (1.22)$$

(ν is the density of states on the Fermi surface; $\Delta/T = \alpha\,(1 - T/T_c)^{1/2}$). In the normal state

$$S^n = \frac{\pi^2}{3}\nu T.$$

We next calculate the specific heats c^s and c^n, and obtain the following expression for the specific heat jump:

$$\beta = \left(\frac{c^s}{c^n}\right)_{T=T_c} = 1 + \frac{3}{2\pi^2}\alpha^2. \quad (1.23)$$

In the weak-coupling approximation $\alpha = 3.06$, which in view of (1.23) leads to $\beta = 2.43$. For Pb (see (1.21) $\alpha = 4$ is obtained. At the same time we find from (1.23) that $\beta_{Pb} = 3.4$, which agrees well with the observed experimental values (see above).

Thus, abandoning the weak coupling approximation results in a higher value for the specific heat jump than in the usual isotropic theory.

A higher value for α helps to explain (see below) the basic feature of the thermal conductivity of strong coupling superconductors. In references /82, 83/ the relationship between $\Delta(0)$ and T_c is calculated. Equation (1.19) is again the starting point in this case. Eliminating the coupling constant g^2 as above and equating the right sides of the equation for Σ_0 at $T = 0$ and $T = T_c$, we arrive at the following equation after lengthy calculations (which we omit here):

$$\Delta(0) = \frac{\pi}{\gamma} T_c \left[1 + 5.3 \frac{T_K^2}{\omega^2} \ln \frac{\omega}{T_c}\right]. \qquad (1.24)$$

If the terms which contain T_c^2/ω^2 are neglected in (1.24), then we arrive at the well-known relationship $\Delta(0) = (\pi/\gamma)\, T_c$ /1/, which is valid for most superconductors, where the electron-phonon interaction is weak and the ratio T_c/Θ is consequently small (Θ is the Debye temperature). In anomalous superconductors the second-order term is not negligible, especially as it is multiplied by a large logarithmic factor.

From equation (1.24) it can be seen, first of all, that the energy gap and T_c are not proportional to each other when the electron-phonon interaction is strong. This fact received experimental confirmation in /99/, where nonproportional changes in $\Delta(0)$ and T_c as a function of the pressure were observed in Pb, Sn, and In films. Moreover, good quantitative agreement with equation (1.24) is noted in /99, 326/.

In comparing equation (1.24) with experimental data, one cannot assume that the characteristic frequency ω is equal to the Debye frequency ω_D. The Debye model is too coarse to describe the properties of strong coupling superconductors. If the phonon spectrum of the superconductor is not distinguished by any specific anomalies, the value of ω should be considered equal to the frequency corresponding to the lower peak of the phonon density of states.

Substituting the experimental data for Pb ($T_c = 7.2°$, $\omega \approx 4.5$ MeV /99/) into (1.24), we find $2\Delta(0) \approx 4.3 T_c$, which agrees fairly well with experimental data establishing the direct relationship between $\Delta(0)$ and T_c (see for example, /5, 6/ and /235/).

8.2. Thermal conductivity

The above equations help to explain the main body of experimental data obtained in the study of the thermodynamic, electromagnetic, and kinetic properties of strong coupling superconductors. In this section we will consider their thermal conductivity. Sound attenuation will be discussed in Chapter III.

To calculate the electronic thermal conductivity of anomalous superconductors (in which electron-phonon interaction is not weak), we will set out from the general definition of the thermal conductivity /19/

$$\varkappa(T, \omega) = (2\omega T)^{-1} \int d^4x \langle [\hat{q}(x), \hat{q}(0)] \rangle \tag{1.25}$$

($\hat{q}$ is the heat flux operator and can be written as /25/

$$\hat{q} = \left(\frac{\nabla\nabla'}{2m} - \mu + V\right) \frac{\nabla - \nabla'}{2mi} \psi^+(x')\psi(x)_{x=x'},$$

where V is the potential energy).

Expression (1.25) can be rearranged:

$$\varkappa = (2\omega)^{-1} \sum_{\omega_{n'}} \int \frac{dp}{(2\pi)^3} \frac{\omega_{n'}\omega'_-}{m^2} p^2 \times L_{\text{av}}, \quad L =$$

$$= \{G^n(p, \omega_{n'}) G^n(-p, -\omega'_-) + F^{n+}(p, \omega_{n'}) F^{n+}(-p, -\omega'_-)\}, \tag{1.25'}$$

$$\omega'_- = \omega_{n'} - \omega_1, \quad \omega_{n'} = (2n'+1)\pi T, \omega_1 = -i\omega.$$

Here G^n and F^{n+} are the Green's functions of the electron excitations interacting with the impurities; G^n and F^{n+} can be written in the form /100/

$$G^n = \frac{i\tilde{\omega} + \xi}{\tilde{\omega}^2 + \xi^2 + \tilde{\Sigma}^2(\omega)}, \qquad F^{n+} = \frac{\tilde{\Sigma}^*(\omega)}{\tilde{\omega}^2 + \xi^2 + \tilde{\Sigma}^2(\omega)},$$

where

$$\tilde{\omega} = \eta\omega, \quad \tilde{\Sigma}(\omega) = \eta\Sigma(\omega), \quad \eta = 1 + \frac{1}{2\tau(\omega^2 + \Sigma^2(\omega))^{1/2}},$$

and τ is the relaxation time.

Substituting the expression for the Green's functions into (1.25') and integrating with the aid of residues, we find that

$$\varkappa(T, \omega) = \frac{\pi N}{2m\omega} \sum_{\omega_{n'}} \omega_{n'}\omega'_- \times \frac{1 - [\omega_{n'}\omega'_- + \Sigma(\omega_{n'})\Sigma(\omega'_-)](\omega^2_{n'} + \Sigma^2(\omega_{n'}))^{-1/2}(\omega'^2_- + \Sigma^2(\omega'_-))^{-1/2}}{(\omega^2_{n'} + \Sigma^2(\omega_{n'}))^{1/2} + (\omega'^2_- + \Sigma^2(\omega'_-))^{1/2} + 1/\tau}.$$

We next replace the summation (see similar calculations in /17/) by integration in the complex variable plane. It is significant that the points which satisfy the equation $\omega = \Sigma(-i\omega)$, that is, the points which are solutions of the equation for the energy gap Δ, are singular. It turns out that the usual steady-state thermal conductivity is described by expressions (1.5) /14/ which, however, contain the energy gap $\Delta(T)$ for the anomalous superconductor (in lead, for example, $\Delta/T \approx 4\,(1 - T/T_c)^{1/2}$) as $T \to T_c$).

In the case of Pb the curve for $\varkappa_{ed}^s\,(T)$, based on (1.5) and (1.21), is given in Figure 15. It can be seen that the $\varkappa_{ed}^s\,(T)$ curve drops more abruptly with decreasing temperature just below the critical point than in the usual case. Such an abrupt fall in $\varkappa_{ed}^s$ for Nb—Mo was observed experimentally by Souca /54/.

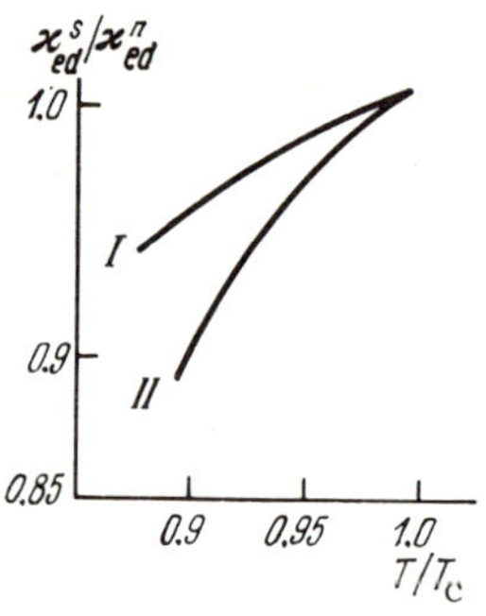

FIGURE 15. $\varkappa_{ed}^s$ as a function of the temperature in strong coupling superconductors. Curve I corresponds to the conventional theory /14/, and curve II is for strong coupling superconductors /85/.

The thermal conductivity of pure anomalous superconductors associated with the scattering of electrons by phonons is described by expression (1.10') with the appropriate function $\Delta\,(T)$. The corresponding theoretical curve (curve 2), which fits the experimental data fairly well /55/, is presented in Figure 13). (Curve I in Figure 13 is constructed according to the BCS theory.)

The larger value of the coefficient α in the equation $(\Delta/T)_{T\to T_c} = \alpha\,(1 - T/T_c)$ determines the more rapid decrease in the thermal conductivity of anomalous superconductors with decreasing temperature than in the usual case.

9. SUPERCONDUCTORS WITH OVERLAPPING ENERGY BANDS

9.1. The Multiband Model

Most superconducting elements have several overlapping unfilled bands, since either the upper unfilled band is not an s-band, or the superconductor is a transition element. Band overlapping is the rule, rather than the exception, and consideration of a multiband structure is necessary for a more detailed comparison of the theory with experiment. The first two-band model was based on the method of (u, v) transformations and developed in papers by Moskalenko /157/ and Suhl, Matthias, and Walker /158/. Consideration of a multiband structure is important in the study of transfer phenomena.

In /96, 160—162/ the properties of superconductors with overlapping bands were studied by the Green's function method. Here both general laws characteristic of the multiband case and the influence of the multiband structure on the various electromagnetic, kinetic, and thermodynamic properties were considered. It is significant that the parameters determined directly by experiment appear in all the fundamental equations in this case.

Several energy gaps were detected experimentally in superconductors with overlapping bands in a series of papers. Data from the observation of several energy gaps in fairly pure specimens of Nb, Ta, and V are given in /152/. For example, for Nb, $2\Delta_1/T_c = 3.5$; $2\Delta_2/T_c = 0.3$ The fine structure of the tunneling characteristics of Nb corresponding to the multiband case was observed in /163/.

In /164/ several energy gaps were noted in Sn, and in /165/ two energy gaps were observed in Pb ($2\Delta_1/T_c = 4.1$; $2\Delta_2/T_c = 1.1$). In /166/ data are presented on the multiband structure in Ga, which is also a strong coupling superconductor; careful measurements in Ga led to the observation of three energy gaps, where $2\Delta_1/T_c = 3.8$, $2\Delta_2/T_c = 4.1$, and $2\Delta_3/T_c = 4.2$. In /167/ two energy gaps were noted in Al, and in /168/ similar results were obtained by ultrasonic measurements in pure Nb and Ta. For a detailed review of the experimental data see /129/.

We shall briefly describe the basic properties of superconductors with overlapping bands. We follow mainly the treatment adopted in /96, 160—162/. First we consider the phenomenon of Cooper pairs in the multiband model.

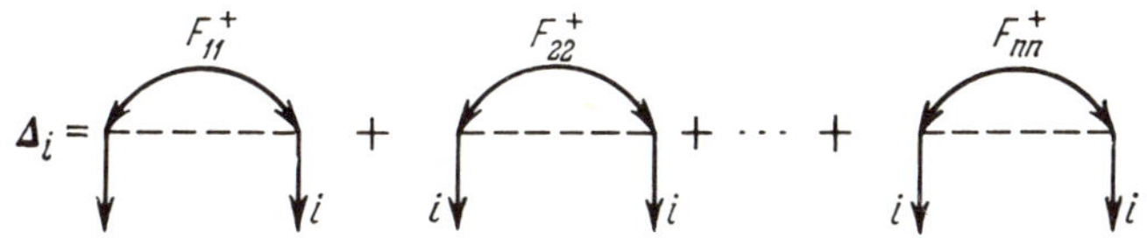

FIGURE 16. Cooper pairs in the multiband model.

When there are several overlapping bands, each is characterized by its own energy gap. The self-energy parts $\Delta_i(\boldsymbol{p}, \omega_n)$, describing the

Cooper pairs in each band satisfy the system of equations in Figure 16 or, in analytical form,

$$\Delta_i(\boldsymbol{p}, \omega_n) = \sum_{l=1}^{n} \sum_{\omega_{n'}} \int \lambda_{il} D(\boldsymbol{p} - \boldsymbol{p}', \omega_n - \omega_{n'}) F_{ll}^{+}(\boldsymbol{p}', \omega_{n'}), \qquad (1.26)$$

where

$$F_{il}^{+} = -i \langle T(\psi_i^{+} \psi_i^{+}) \rangle \text{ [3]}, \quad F_{ll}^{+}(\boldsymbol{p}, \omega_n) = \Delta_l/(\omega_n^2 + \xi^2 + \Delta_l^2);$$

D is the phonon Green's function; λ_{il} describes the transition of an electron pair from the i-th to the l-th band with absorption or emission of a phonon. We do not consider the pairing of electrons occupying different bands (which involves neglecting the terms with the off-diagonal elements F_{il}^{+} $(i \neq l)$), since overlap is unlikely for electrons with opposite spins near the Fermi surface.

Thus, each band has its own condensate of Cooper pairs. This does not necessarily mean that the pairs in each band form completely independently. Cooper pairs are formed in a particular band first due to the attractive forces between electrons within the band and second due to transitions of pairs of particles from other bands. These transitions are described by the off-diagonal elements in (1.26).

Introducing the effective coupling constants in the usual way and summing over $\omega_{n'}$, we write (1.26) in the form

$$\Delta_i = \sum_{l=1}^{n} g_{il} \nu_l \Delta_l \int d\xi_l \frac{\operatorname{th}(\varepsilon_l/2T)}{\varepsilon_l}, \qquad (1.27)$$

$$\varepsilon_l = (\xi_l^2 + |\Delta_l|^2)^{1/2}.$$

For $T = 0$,

$$\Delta_i = \sum_l g_{il} \nu_l \Delta_l \ln \frac{2\tilde{\omega}}{|\Delta_l|}; \qquad (1.27')$$

and for $T = T_c$,

$$\Delta_i = \ln \frac{1.14\tilde{\omega}}{T_c} \sum_l g_{il} \nu_l \Delta_l. \qquad (1.27'').$$

For simplicity we assume that $\xi_l = p^2/2m_l - \mu$, m_l is the effective mass, ν_l is the density of states in the l-th band, $\nu_l = p_{Fl}\, m_l\, (2\pi^2)^{-1}$. It is easy to show that $g_{il} = g_{li}^{*}$. The interaction, as usual, is cut off at the frequency $\tilde{\omega} \sim \omega_D$.

It can be seen in (1.27) that the presence of several bands favors pair formation. For example, eliminating Δ_2, we obtain for two bands

$$\Delta_1 = \ln \frac{1.14\tilde{\omega}}{T_c}\left[g_{11}\nu_1 + \frac{|g_{12}|^2 \nu_1\nu_2}{[\ln(1.14\tilde{\omega}/T_c)]^{-1} - g_{22}\nu_2}\right]\Delta_1.$$

Thus, whereas pair formation is determined by the factor $g_{11}\nu_1$, in the single-band case, in the two-band case this factor is incremented by the positive (regardless of the sign of g_{12}, even in the case of repulsion) term

$$\frac{|g_{12}|^2 \nu_1\nu_2}{[\ln(1.14\tilde{\omega}/T_c)]^{-1} - g_{22}\nu_2}\left[\ln \frac{1.14\tilde{\omega}}{T_c}\right]^{-1} > g_{22}\nu_2.$$

From the system of equations (1.27) it can be seen that the energy gaps of the different bands are related by a linear function; therefore, they all vanish at one temperature, T_c, which can easily be calculated from (1.27):

$$T_c = 1.14\tilde{\omega} \exp\left(-1\Big/\sum_l \rho_{1l}\eta_l\right),$$

where

$$\rho_{il} = g_{il}\nu_l, \quad \eta_{lc} = (\Delta_l/\Delta_1)_{T=T_c}.$$

The quantities η_{lc} are also determined from the system of equations (1.27). If α is the smallest eigenvalue of the homogeneous system of equations (1.27), that is

$$\varphi_i = \alpha \sum_l g_{il}\nu_l\varphi_l,$$

then

$$T_c = 1.14\tilde{\omega}e^{-\alpha}.$$

In the two-band model the critical temperature is

$$T_c = 1.14\tilde{\omega} \exp\left(-\frac{1}{\rho_{11} + \rho_{12}\eta_{2c}}\right).$$

Calculating $\eta_{2c} = (\Delta_2/\Delta_1)_{T=T_c}$, we obtain

$$T_c = 1.14\tilde{\omega}\exp(-1/\rho), \tag{1.28}$$

where

$$\rho = {}^1/_2\,[\rho_{11} + \rho_{22} + \sqrt{(\rho_{22} - \rho_{11})^2 + 4\rho_{12}\rho_{21}}].$$

To solve the system of equations (1.26) in the weak-coupling approximation, Pokrovskii's method can be used (for further details see Chapter III, Sec. 2). We seek a solution in the form $\Delta_i = 2\tilde{\omega}\varphi_i e^{-\alpha}$. In the zeroth approximation, corresponding to the weak-coupling approximation (for further details, see /160/), we obtain

$$\varphi_i = \alpha \sum_l \rho_{il}\varphi_l. \tag{1.29}$$

Using the exact equation leads to the following solution:

$$\varphi_i = Q\chi_i,$$

where χ_i is the normalized solution of the homogeneous system of linear equations (1.29) corresponding to the smallest eigenvalue α, and

$$Q = \exp\left[-\left(\sum_i \nu_i |\chi_i|^2 \ln|\chi_i|\right) \sum_i (\nu_i |\chi_i|^2)^{-1}\right].$$

The equality of the ratio Δ_i/Δ_1 at $T = 0$ and $T \to T_c$ ($\eta_0 = \eta_{ic}$) follows from equations (1.27'') and (1.29) /28/. Similarly, $\eta_{2c} = (\Delta_2/\Delta_1)_{T\to T_c}$ (for simplicity we limit the discussion to the two-band model) is determined, according to (1.27''), by solving the equation

$$\rho_{12}\eta_{2c}^2 + (\rho_{11} - \rho_{12})\,\eta_{2c} + \rho_{21} = 0.$$

This equation enables us to express η_{2c} in terms of the effective coupling constants ρ_{il}. The ratio $\eta_{2\,0}$ in the weak-coupling approximation can be determined with the aid of system of equations (1.29). It is easy to see that $\eta_{2\,0}$ is expressed in terms of the effective coupling constants ρ_{il} by the same relationship as η_{2c}. In both cases we seek a solution of the homogeneous system of equations which corresponds to the same smallest eigenvalue.

Thus, we conclude that the ratios $\eta_2 = \Delta_2/\Delta_1$ at $T = 0$ and at $T \to T_c$ are indeed equal. This condition is essential in analyzing the experimental data and especially in determining multiband superconductor parameters.

In the usual theory the critical temperature, T_c, is the parameter which determines the special properties of the superconducting state. The energy gap $\Delta(0)$ is related to T_c in the weak-coupling approximation by the universal relationship $\Delta(0) = (\pi/\gamma)\,T_c$. In the multiband model, which is characterized by a large number of effective coupling constants, no such universal relationship between $\Delta_i(0)$ ($i = 1, 2, \ldots, n$) and T_c is observed in the general case.

and the system must be described by other independent parameters besides T_c. These additional parameters must be determined experimentally. They are in particular the values of $\Delta_i(0)$ and $(\Delta_1)_{T\to T_c}$. The other energy gaps $(\Delta_i)_{T\to T_c}$ are easily calculated from the equality $\eta_0 = \eta_c$. For a more detailed discussion of the determination of the parameters for multiband superconductors, see Chapter III, Sec. 4.

The values of all the fundamental thermodynamic, electromagnetic, and kinetic quantities can be expressed in terms of the parameters Δ_1, T_c, and η_c, which are determined experimentally (e. g. by tunneling measurements). For example, as $T\to 0$ the thermodynamic critical field H_c is equal to

$$H_c(T) = H_c(0)\left[1 - \frac{\pi^2}{3\chi_1^2}\frac{\sum_i \nu_i}{\sum_i \nu_i \eta_i^2(0)}\frac{T^2}{T_c^2}\right],$$

where $\chi_1 = \Delta_1(0)/T_c$, $\eta_i(0) = \Delta_i(0)/\Delta_1(0)$, ν_i is the normal density of states in the i-th band. The specific heat as $T\to 0$ is expressed by the equation

$$\frac{c^s(T)}{c^n(T_c)} = \frac{3\sqrt{2}}{\pi^{3/2}}\sum_{i=1}^{n}\nu_i\frac{\Delta_i(0)}{T_c}\left(\frac{\Delta_i(0)}{T}\right)^{3/2}\exp\left(-\frac{\Delta_i(0)}{T}\right)\left(\sum_i \nu_i\right)^{-1}.$$

Finally, we mention a very important property of the multiband model. Because of the scattering by impurities the wave function of each electron is a linear combination of the wave functions of different bands, i. e., scattering mixes the states of the different bands. As a result, for sufficiently doped superconductors, i. e., for $l \lesssim v_F/\Delta_{av}$, the superconductor behaves as a single-band system. Therefore, all the relationships obtained below are valid only for pure multiband superconductors, i. e., for the case $l \gg v_F/\Delta_{av}$ (l is the free path).

The influence of impurities on the behavior of a two-band superconductor was studied in detail in /169, 170, 320/. Note that if the energy gap is defined as the absorption threshold (i. e., the frequency at which the imaginary part of the Green's function becomes different from zero), then at the frequency $\omega = \Delta_1$ (Δ_1 is the minimum energy gap) the electrons of the other bands, strictly speaking, will also contribute to the absorption due to interband transitions caused by impurities. In this sense, the introduction of a very small quantity of impurity isotropizes the energy gap /130/ and leads to a state with one energy gap in the multiband case /320/. However, the corresponding effect will be small, and for low concentrations

($l \gg \xi_0$) the density of states, which determines the absorption, will have maxima at the frequencies $\omega_i \approx \Delta_i$ (Δ_i are the energy gaps in the pure superconductor). Therefore, at a low impurity concentration, we can consider the energy gap to be anisotropic (see Chapter III, Sec. 9) or to possess a multiband structure. Only if $l \lesssim \xi_0$ is the energy gap completely isotropic. Then the absorption maxima begin to coincide in the multiband case, and consequently one energy gap appears in the specific heat and other quantities.

It is significant that the effects associated with the multiband structure (and with anisotropy) are easily isolated experimentally, since the special features of the temperature dependence should disappear when the superconductor is doped to the impurity concentration given by the condition $l \lesssim \xi_0$. This is confirmed in experiments by the measurement of the specific heat c_V of pure Pb and Pb doped with In at low temperatures /171/. At 1.76 at. % In, the specific heat falls more rapidly than for pure Pb with decreasing temperature (while the behavior of c_V as $T \to 0$ is determined by the smallest energy gap $\Delta_{\min}$; $\Delta_{\min} < 1.76\ T_c$), and at 5.9 at. % In the fall becomes even faster.

It should be noted that the relationship for Pb from the strong-coupling theory, $2\Delta(0) = 4.1\ T_c$, is valid for quite dirty Pb, since in Pb there are two /165/ or three /35/ overlapping bands. The experimental data obtained, for example, from the study of the absorption of electromagnetic waves (which give $\Delta(0) \approx 4.1\ T_c$) were apparently obtained for quite dirty lead, since only one energy gap was observed, but in subsequent data on specific heat for $T \to 0$ two energy gaps were observed.

We indicated above that a multiband structure favors pair formation and increases the effective coupling constant. On the other hand, impurity atoms cause (through interband transitions) a weakening and subsequent disappearance of the multiband effects. Therefore, the introduction of impurities should lead to a weakening of the effective attraction and a decrease in T_c, which is experimentally observed. This effect is absent in the single-band isotropic case /100, 172/. The latter is characterized by a T_c which is independent of the concentration of nonmagnetic impurities. Here, as before, with the anisotropic model we naturally do not take into account the influence of impurities on the phonon spectrum of the metal.

We have mentioned that the anisotropy effect results in comparatively small changes in the energy gap as a function of orientation. In the presence of overlapping bands, however, the difference between $\Delta_{\max}$ and $\Delta_{\min}$ can be very large. The energy gaps of different bands can differ several-fold, and this fact also makes consideration of the multiband effects important.

9.2. Electronic Thermal Conductivity

The multiband effects lead to a number of peculiarities in the temperature dependence of the kinetic coefficients /28/. Appropriate measurements can be used to determine the parameters which characterize superconductors with overlapping bands.

Since the effects associated with the presence of several energy gaps emerge only in fairly pure superconductors ($l \gg \xi_0$), the electronic component evidently makes the major contribution to the heat flux in the temperature range close to the critical point. The phonon conductivity is substantial in the intermediate temperature range and when $T \to 0$.

We first consider the properties of high-purity specimens, in which the scattering of electrons by phonons is the major factor.

Scattering of electrons by phonons. In this case, the total electronic heat flux is composed of the heat fluxes created by the various bands: $Q = \sum_i Q_i$. The Hamiltonian of the system has the form

$$H = \sum_{\substack{k,\, s,\, q,\\ k'=k+q}} V^i_{kk'}\, \overset{i+}{a}_{ks} \overset{i}{a}_{k's} b^+_q + \text{h.c.}$$

(i is the band index). We do not consider processes in which the electrons move from one band to another upon being scattered by a thermal phonon, since such transitions are usually accompanied by a change $\sim p_F$ in the electron momentum, which is impossible at the thermal phonon frequencies $\omega \sim T$. Interband transitions are an important factor in the scattering of electrons by impurities and must be considered in calculating $\varkappa^s_{ed}$.

We next turn to the new Fermi operators,

$$\overset{i}{\alpha}_{k0} = \overset{i}{u}_k \overset{i}{a}_{k,\,1/2} - \overset{i}{v}_k \overset{+}{a}_{-k,\,-1/2}, \quad \overset{i}{\alpha}_{k1} = \overset{i}{u}_k a_{-k,\,-1/2} + \overset{i}{v}_k \overset{+}{a}_{k,\,1/2}, \tag{1.30}$$

$$u^2_k = {}^1/_2\,(1 + \xi/\varepsilon), \quad v^2_k = {}^1/_2\,(1 - \xi/\varepsilon).$$

The kinetic equation for the electronic distribution function of the i-th band has the form

$$\frac{\partial f^i}{\partial x}\,\frac{\partial \varepsilon^i}{\partial p_x} - \frac{\partial f^i}{\partial p_x}\,\frac{\partial \varepsilon^i}{\partial x} = \left(\frac{\partial f^i}{\partial t}\right)_{\text{col}}. \tag{1.31}$$

We seek a solution in the form

$$f^i = f^i_0 + \frac{\partial f^i_0}{\partial \varepsilon}\,\varphi^i(\varepsilon,\ \Omega)\,\frac{\partial T}{\partial x}\,.$$

The collision integral has the same form as in the single-band case (see equations (1.8) and (1.8')) except that it contains the functions f^i and φ^i, which characterize the electronic distribution function in the i-th band.

The kinetic equation is solved by the method of Landau and Pomeranchuk (see Sec. 3). It is significant that integration over the polar angle ϑ (we first integrate over ξ and then over the excitation energy ε) gives rise to the factor $m^i/(p^i q)$. Next, as in the single-band case, we write $(p_x/p)^i=(p_x/p)_{p=p^i_F}+a^i_x\xi$, where $a^i_x\sim p^i_F/\varepsilon^i_F$ (ε^i_F is the Fermi level reckoned from the bottom of the i-th band). The solution is next sought in the form

$$\varphi^i(\varepsilon, \Omega, T)=\varphi^i_1(\Omega, T)+\varphi^i_2(\varepsilon, \Omega, T), \quad \varphi_1\gg\varphi_2.$$

Finally we arrive at the following expression for the function $\varphi^i_1(\Omega, T)$:

$$\varphi^i_1=\frac{\lambda^i A_\Omega}{T^4\Phi^i(T)}\,g^i(T). \tag{1.32}$$

Here

$$g^i(T)=\int\limits_{b^i}^{\infty} f_0^2 e^z z(z^2-b^{i2})dz,$$

$$\Phi^i(T)=\int\limits_0^\infty\frac{4x^4dx}{e^x-1}\int\limits_{b^i}^\infty\frac{dz}{(e^z+1)(e^{-z-x}+1)}+\int\limits_{2b^i}^\infty\frac{x^4dx}{e^x-1}\int\limits_{b^i}^{x-b^i}\frac{dz}{(e^z+1)(e^{-z}+e^{-x})},$$

$b^i=\Delta^i/T$, A_Ω is a function of the angles. The coefficient λ^i is $p^{i4}_F/\varepsilon^i_F m^{i2}I^{i2}\sim\varepsilon^i_F/I^{i2}$; the factors p^i_F/m^i and p^i_F/ε^i_F arise when integrating over ϑ and substituting the coefficient a^i_x, respectively; the appearance of p^{i2}_F in the expression for λ^i is associated with the expansion of the function φ^i in a series in powers of q/p^i_F up to second-order terms (see Sec. 3); $I^i=|v^i_F|/q$.

The matrix element of the interaction of electrons with low-energy phonons, as we know, can be written in the form (see, for example, /35/) $V_{kk'}\sim q^{1/2}I_{kk'}$ ($I_{kk'}$ has the dimensions of energy). This function can be calculated only in certain models. A treatment based on the adiabatic theory /97, 319/ leads to the conclusion that the "rigid" ion approximation is valid. The value of $I^i_{kk'}$ calculated by the method of Wigner and Seitz (see, for example, /35/) equals $I^i=(n^i/\nu^i)F=\varepsilon^i_F F$ (n^i is the electron concentration; ν^i is the density of states; $\nu^i\sim m^i p^i_F$ is the effective mass approximation; F is expressed through the overlap integral).

The coefficients λ^i are thus determined by the parameters which characterize the normal state of each band.

We next calculate the heat flux by means of the formula $Q = (4\pi^3)^{-1}\sum_i \int \varepsilon v_x f^i \frac{dp^i}{(2\pi)^3}$ and obtain:

$$Q=(4\pi^3)^{-1}\sum_i \int \varepsilon v_x \frac{\partial f_0}{\partial \varepsilon}\cdot \frac{1}{T}\varphi^i(\Omega, T)\,\nu^i d\xi\, d\Omega \frac{\partial T}{\partial x}, \tag{1.33}$$

where $v_x = \partial\varepsilon/\partial p_x = \xi p_x^i/(\varepsilon m^i)$; the function φ^i is determined by equation (1.32), and ν^i is the density of states in the i-th band. We substitute v_x, φ^i , and ν^i into the expression for Q. Next expanding p_x^i in powers of ξ/ε_F^i and integrating over the excitation energy, we arrive, after simple transformations, at the following final expression for the ratio of the thermal conductivities in the superconducting and normal states, $\varkappa_{ep}^s/\varkappa_{ep}^n$:

$$\varkappa_{ep}^s/\varkappa_{ep}^n = \sum_i \rho^i G(b^i), \tag{1.34}$$

where

$$\rho^i = B^i \Big/ \Big(\sum_i G(T_\text{K})\, B^i\Big),$$

$$B^i = \lambda^i p_F^{i2}/\varepsilon_F^i = \nu^i p_F^i / I^{i2},\ G(b^i) = g^2(b^i)/\Phi(b^i),$$

and the functions $g(b)$ and $\Phi(b)$ have the same form as in the single-band case and are determined from (1.32) and (3.41).

The relative contribution of each band is determined not only by the energy gaps Δ^i , but also by the parameters ρ^i , which are strongly dependent on the λ^i, describing the electron-phonon interaction for each band. For example, in the case of transition metals which have overlapping s and d bands $(\nu^d \gg \nu^s)$, the contribution of the d-band to the thermal conductivity may in general surpass the contribution of the s-electrons, despite the latter's greater mobility. This can be explained in the following way. Because of the weaker interaction of the d-electrons with the phonons, the distribution function of the d-electrons is more appreciably perturbed by the temperature gradient, which produces a larger contribution to the heat flux (1.33). Similar reasoning should also be applied in considering the conductivity of pure transition metals in the normal state.

In the temperature range near the critical point, where this mechanism is the major factor in pure superconductors, the function $G(b^i)$ equals

$$G(b^i) = \frac{1}{\Phi^i(T)}\left[2\sum_{s=1}^{\infty}\frac{(-1)^{s+1}}{s^2}e^{-b^i s} + 2b^i \ln(1+e^{-b^i}) - \frac{b^{i2}}{2(e^{b^i}+1)}\right]^2. \tag{1.34'}$$

The thermal conductivity falls exponentially with decreasing temperature, but more slowly than in the single-band case.

Scattering of electrons by impurities. The calculation of $\varkappa_{ed}^{s}$ in the multiband case is meaningful only for $l \gg \xi_0$. If the concentration is such that $l \lesssim \xi_0$, the energy gaps of the different bands are equal and the thermal conductivity is described by equations (1.5) and (1.14). However, if the impurity concentration is low, it is not a priori clear whether scattering off impurities and defects makes a substantial contribution or whether the scattering of electrons by phonons is the major factor. Thus the range in which the following treatment can be applied has to be established.

The situation in which $\varkappa_{ed}^{s}$ is the major factor and yet the multiband effects are substantial accords with reality. Experimental data obtained, for example, in /15, 61/ show that the scattering of electrons by phonons is substantial only in specimens of high purity, when $\varkappa_{ed}^{s}$ is entirely determined by the scattering on the specimen's boundaries. When the free path is about 10^{-2} cm, the contribution of the impurity at $T \approx T_c$ becomes comparable to the contribution of the electron-phonon scattering. Thus, for example, for tin the corresponding free path is $l = 1.75 \cdot 10^{-2}$ cm /15/, for thallium $l = 6.5 \cdot 10^{-3}$ cm /15/, and for indium $l \approx 10^{-2}$ cm.

At high concentrations $\varkappa_{ed}^{s}$ begins to be the major factor. Thus, there is an impurity concentration range ($10^{-1}\,\text{cm} \gg l \gg \xi_0 \sim 10^{-4}\,\text{cm}$) in which, on the one hand, the multiband effects are substantial and, on the other hand, the scattering of electrons by phonons is a major factor. This is most often true in superconductors in which the ratio T_c/Θ is small (we remember that the electronic component of the thermal conductivity is dominant near T_c), since the free path $l_{ep} \sim (\Theta/T)^3$, and therefore, when the impurity concentration is low, the inequality $l_{ed} \lesssim l_{ep}$ is most easily fulfilled in superconductors with a low T_c/Θ. These are the usual weak coupling superconductors. Such substances were also studied in /15/.

We next proceed to calculate $\varkappa_{ed}^{s}$ /28/. The Hamiltonian of this system has the form

$$H = \sum_{k,k',s,i} V_{kk'}^{ii} a_{k's}^{i+} a_{ks}^{i} + \sum_{\substack{i,l \\ i \neq l}} V_{kk'}^{il} a_{k's}^{i+} a_{ks}^{l}.$$

In calculating $\varkappa_{ed}^{s}$, interband electronic transitions caused by elastic collisions between electrons and impurity atoms must be considered. In the final analysis, with an increase in the impurity concentration, these transitions cause mixing of the wave functions of different bands and disappearance of the multiband effects.

Next we must solve the system of kinetic equations for the distribution functions corresponding to the various bands, at the same time allowing for the interband transitions in the collision integrals. However, we will not go beyond the most important special case of the overlapping of the s- and d-bands, and we will assume that the inequalities $\nu^d \gg \nu^s$, $v^s \gg v^d$ are fulfilled. Under these conditions we can assume, as in the conduction theory of transition metals (see, for example, /35/), that the heat flux is transferred entirely by the s-electrons. The basis for this assumption will be presented below. We note that it is valid only for $\varkappa_{ed}^{s}$.

In /29/ the coefficient $\varkappa_{ed}^{s}$ is calculated in the multiband case, on the assumption that the interband transitions are the major component of the thermal conductivity of the s-band (since $\nu^s \ll \nu^d$) and the contribution of the electronic transitions within each band is neglected. However, the inequality $V^{ss} \gg V^{sd}$ increases the contribution of the intraband scattering. Therefore, the approximation adopted in /29/ is valid if $\nu^d \gg \nu^s (V^{ss}/V^{sd})^2$. Moreover, in carrying out the numerical calculations based on /16/, it is assumed in /29/ that $\Delta^s/\Delta^d \approx 0$.

We write the kinetic equation for the s-electrons, and change over as usual to the new Fermi operators (1.30):

$$\frac{\partial f_0^s}{\partial \varepsilon}\frac{\partial \varepsilon}{\partial p_x}\frac{\varepsilon}{T}\frac{\partial T}{\partial x} = -\int |V_{kk'}^{ss}|^2 (u_k^s u_{k'}^s - v_k^s v_{k'}^s)^2 \times$$

$$\times (f_k^s - f_{k'}^s)\,\delta(\varepsilon_k - \varepsilon_{k'})\,dk' + \int |V_{kk'}^{sd}|^2 (u_k^s u_{k'}^d - v_k^s v_{k'}^d)^2 \times (f_k^s - f_{k'}^d)\,\delta(\varepsilon_k - \varepsilon_{k'})\,dk'.$$

A similar equation can also be written for the d-electrons. Substituting the values of $u_{k'}^{s,d}$ and $v_{k'}^{s,d}$, we arrive at the following equation after some simple rearrangements:

$$-\frac{\partial f_0^s}{\partial \varepsilon}\frac{\partial \varepsilon}{\partial p_x}\frac{\varepsilon}{T}\frac{\partial T}{\partial x} = -\frac{|\xi_k^s|}{\varepsilon_k}\int |V_{kk'}^{ss}|^2 (f_k^s - f_{k'}^s)\,\nu^s do -$$

$$-\left(1 + \frac{\xi_k^s \xi_k^d - \Delta^s \Delta^d}{\varepsilon_k^2}\right)\frac{\varepsilon_k}{|\xi_k^d|}\int |V_{kk'}^{sd}|^2 (f_k^s - f_{k'}^d)\,\nu^d do. \qquad (1.35)$$

Here $\xi_k^{s,d} = (\varepsilon_k^2 - \Delta^{2s,d})^{1/2}$.

The distribution function of the electron excitations is sought as usual in the form

$$f = f_0^i + f'^i,\ f'^i = \frac{\partial f_0}{\partial \varepsilon}\, c^i(\varepsilon) v_x^i \frac{\partial T}{\partial x}.$$

We next assume in solving (1.35) that $f'^d \ll f'^s$. Considering both equation (1.35) and the corresponding kinetic equation for the d-electrons, one can easily prove the validity of this approximation. It turns out that

$$f'^d/f'^s \sim \left[\frac{\tau_d^n}{\tau_s^n}\frac{v_x^d}{v_x^s} + \frac{\tau_d^n}{\tau_{ds}^n}\right],$$

where

$$(\tau_{il}^{n})^{-1} = \int |V^{il}|^2 \nu^l do;$$

$$(\tau_{d}^{n})^{-1} = (\tau_{dd}^{n})^{-1} + (\tau_{ds}^{n})^{-1};$$

$$(\tau_{s}^{n})^{-1} = (\tau_{ss}^{n})^{-1} + (\tau_{sd}^{n})^{-1};$$

$$(\tau_{ii}^{n})^{-1} = \int |V^{ii}|^2 (1 - \cos\theta)\, \nu^i do.$$

These values are expressed in terms of the parameters of normal metals. It is easy to see that f'^d is in fact much smaller than f'^s (since $v_x^d \ll v_x^s$, $\tau_d^n \ll \tau_{ds}^n$; $|V^{il}|^2 \ll |V^{ii}|^2$, which can be shown by evaluating the matrix elements for the scattering on a charged, screened impurity).

In this approximation, we can write equation (1.35) in the form

$$-\frac{\partial f_0^s}{\partial \varepsilon}\frac{\partial \varepsilon}{\partial p_x}\frac{\varepsilon}{T}\frac{\partial T}{\partial x} = -\frac{f^s - f_0^s}{\tau}, \tag{1.36}$$

$$\tau^{-1} = \tau_{ss}^{-1} + \tau_{sd}^{-1}, \qquad \tau_{ss}^{-1} = \frac{|\xi^s|}{\varepsilon_k}(\tau_{ss}^{n})^{-1},$$

$$\tau_{sd}^{-1} = \frac{1}{2}\left(1 + \frac{\xi^s\xi^d - \Delta^s\Delta^d}{\varepsilon_k^2}\right)\frac{\varepsilon_k}{|\xi_k^d|}(\tau_{sd}^{n})^{-1},$$

where τ_{il}^n is the corresponding relaxation time for the normal metal. We next find the perturbed distribution function $f'^s = f^s - f_0^s$ from (1.36) and calculate the heat flux from $Q \sim \int \varepsilon v_x f'^s d\boldsymbol{p}$. In the temperature range close to the critical point, the thermal conductivity is

$$\varkappa_{ed}^s(T) = \frac{\varkappa_{ed}^n(T_c)}{F(T_c)}\left[F(T) - \frac{(\Delta^s - \Delta^d)^2}{2T}\frac{\tau^n}{\tau_{sd}^n}\right], \tag{1.37}$$

$$\varkappa_{ed}^n(T_c) = \frac{2}{3}\frac{p_F^{s^3} F(T_c)}{\pi^2 m^s}\tau^n,$$

where $\tau^n = (\tau_{ss}^n)^{-1} + (\tau_{sd}^n)^{-1}$, and the function $F(T)$ is determined by equality (1.5).

Scattering off impurities, which is accompanied by interband transitions, is an additional relaxation mechanism. The existence of this mechanism naturally leads to a smaller relaxation time than in the usual case and accordingly to a decrease in the thermal conductivity. This additional decrease is described by the second term in equation (1.37).

9.3. Lattice Conductivity

At temperatures not too close to T_c, the heat flux in the lattice begins to have an effect. The important quantity is $\varkappa_{pe}^s$ in the intermediate temperature range and $\varkappa_{pd}^s$ when $T \to 0$. Investigation of $\varkappa_{pe}^s$ enables us to determine the energy gaps in the various bands.

The collision integral which appears on the right in the kinetic equation (1.11) for the phonon distribution function can be written in the form $(\partial N/\partial t)_{\text{col}} = \sum_i (\partial N/\partial t)_{\text{col}}^i$, where $(\partial N/\partial t)_{\text{col}}^i$ is the collision integral, describing the scattering of phonons by electrons of the i-th band:

$$\left(\frac{\partial N}{\partial t}\right)_{\text{col}}^i = \int |V_{kk'}^i|^2 \left\{\left(1 + \frac{\xi\xi' - \Delta^{i2}}{\varepsilon^i \varepsilon^{i'}}\right) \times \right.$$

$$\times [f'(1-f)(N+1) - Nf(1-f')]^i \,\delta(\varepsilon' - \varepsilon - \omega) +$$

$$\left. + \frac{1}{2}\left(1 + \frac{\xi\xi' - \Delta^{i2}}{\varepsilon^i \varepsilon^{i'}}\right)[(N+1)ff' - N(1-f)(1-f')]^i \times \delta(\varepsilon' + \varepsilon - \omega)\right\} v^i d\xi\, do.$$

As in the calculation of $\varkappa_{ep}^s$, we do not consider processes in which electrons pass from one band to another.

The kinetic equation is next solved as in the single-band case (see Sec. 4), and the following expression is obtained for the increment R to the phonon distribution function:

$$R = \frac{1}{T^3}\left(\sum_i m^{i2} |V^{i'}|_F^2 \varphi(b^i, x)\right)^{-1} q_x N_0^2 e^x;$$

$$b^i = \Delta^i/T, \quad x = \omega/T,$$

where

$$\varphi(b^i, x) \approx 2x - 2\ln\frac{e^{b^i + x} + 1}{e^{b^i} + 1} +$$

$$+ D^i(x)\left(2b^i - x + 2\ln\frac{e^{x - b^i} + 1}{e^{b^i} + 1}\right),$$

$$D^i(x) = \begin{cases} 1 & \text{for } x \geqslant 2b^i, \\ 0 & \text{for } x < 2b^i. \end{cases}$$

We next calculate the heat flux in the lattice. The ratio $\varkappa_p^s/\varkappa_p^n$ turns out to be

$$\frac{\varkappa_p^s}{\varkappa_p^n} = \frac{I}{I^n} = \frac{\displaystyle\int_0^\infty \frac{x^4 e^x}{(e^x - 1)^2}\left(\sum_i m^i |V^{i'}|_F^2 \varphi(b^i, x)\right)^{-1} dx}{\displaystyle\int_0^\infty \frac{x^4 e^x}{(e^x - 1)^2 \varphi(x)}\left(\sum_i m^i |V^{i'}|_F^2\right)^{-1} dx}. \tag{1.38}$$

In calculating the integral in the numerator of (1.38), it is necessary to consider the step functions $D^i(x)$, which describe the corresponding threshold effects. In superconductors with overlapping bands, each of these functions has its own threshold frequency. Thus, there is always a range of frequencies in which one of the bands is characterized by scattering processes only, while in the other band the creation of excitation pairs by phonons must also be considered. Furthermore, the integral I in the case with two overlapping bands can be written in the form

$$I = I_1 + I_2 + I_3,$$

where (do not confuse the band index 2 with raising to the second power!)

$$I_1 = \sum_i a^i \int_0^{2b^1} \frac{x^4 e^x}{(e^x - 1)^2}\left(x - \ln\frac{e^{b^i+x}+1}{e^{b^i}+1}\right),$$

$$I_2 = \int_{2b^1}^{2b^2} \frac{x^4 e^x}{(e^x - 1)^2}\left[a^1\left(x + 2b^1 + 2\ln\frac{e^{b^1+x}+1}{e^{x-b^1}+1}\right) + \right.$$

$$\left. + a^2\left(x - \ln\frac{e^{b^2+x}+1}{e^{b^2}+1}\right)\right]^{-1},$$

$$I_3 = \sum_i a^i \int_{2b^2}^{\infty} \frac{x^4 e^x}{(e^x - 1)^2}\left(x + 2b^i + 2\ln\frac{e^{b^i+x}+1}{e^{x-b^i}+1}\right),$$

$$a^i = (m^i \,|\, V^{i'} |)^{-2}.$$

Evaluation of the integrals leads to the following final result:

$$\varkappa^s = \varkappa^n I(T)/I(T_c), \tag{1.39}$$

$$I(T) = \frac{a^1 e^{-2b^1} + a^2 e^{-2b^2}}{(a^1 e^{-b^1} + a^2 e^{-b^2})^2} f(b^1) + \frac{a^2 e^{-b^2}(1 - e^{b^1-b^2})}{(a^1 e^{-b^1} + a^2 e^{-b^2})^2} g(b^1) +$$

$$+ \frac{1}{a^1 + a^2}\varphi(b^2, \lambda) + \frac{1}{a^1}[\varphi(b^1, \rho) - \varphi(b^2, \rho)], \tag{1.40}$$

where

$$f(b) = -8(b^4 + b^3)(e^b - 1)^{-1} + 6\zeta(3)(e^b + 1) - 3(e^b + 1) \times$$

$$\times \sum_{s=1}^{\infty} s^{-3} e^{-2bs}(4b^2 s^2 + 4bs + 2) + 6\zeta(4)(e^b - 1) + (e^b - 1) \times$$

$$\times \sum_{s=1}^{\infty} s^{-4} e^{-2bs}(8b^3 s^3 + 12b^2 s^2 + 12bs + 6),$$

$$g(b) = -8(b^4 + b^3)(e^{2b} - 1)^{-1} - 4b^4(e^{2b} - 1)^{-2} - 6\zeta(4) -$$

$$- \sum_{s=1}^{\infty} s^{-4} e^{-2bs}(8b^3 s^3 + 12b^2 s^2 + 12bs + 6) + 6\zeta(3) -$$

$$- 3\sum_{s=1}^{\infty} s^{-3} e^{-2bs}(4b^2 s^2 + 4bs + 2),$$

$$\varphi(z, y) = 8\frac{(z+y)(z^2+y^2)}{e^{2z}-1} -$$

$$-A^4\sum_{s=1}^{\infty} s\exp(-2zs)\,\mathrm{Ei}\,[-s(2z-A)] + 6\sum_{s=1}^{\infty} s^{-3}\exp(-2zs),$$

$$A \approx 2b,\quad \lambda = \frac{a^1b^1+a^2b^2}{a^1+a^2},\quad \rho = b^1 + \frac{a^2}{a^1}e^{-b^2}.$$

To fix ideas, let $b^2 > b^1$.

This treatment can be generalized without difficulty to the case where the number of bands n is greater than 2. The function $I(T)$, which determines the ratio $\varkappa^s/\varkappa^n$ according to equation (1.39), now becomes

$$I(T) = \sum_i a^i e^{-2b^i}\left[\frac{1}{n}\left(\sum_i a^i e^{-b^i}\right)^2\right]^{-1} f(b^1) +$$
$$+\sum_i a^i e^{-b^i}(1-e^{b^1-b^i})\left[\frac{1}{n}\left(\sum_i a^i e^{-b^i}\right)^2\right]^{-1} g(b^1) +$$
$$+\left(\sum_i a^i\right)^{-1}\varphi(b^n, \lambda^n) +$$
$$+\sum_{m=1}^{n-1}\left(\sum_{i=1}^{m} a^i\right)^{-1}[\varphi(b^m, \rho^m) - \varphi(b^{m+1}, \rho^m)], \qquad (1.40')$$

$$A^i \approx 2b^i,\ \lambda^n = \sum_{i=1}^{n} a^i b^i\left(\sum_{i=1}^{n} a^i\right)^{-1},$$

$$\rho^m = \frac{1}{2}\sum_{i=1}^{m-1} A^i + \left(2\sum_{i=1}^{m} a^i\right)^{-1}\sum_{k=m+1}^{n} a^k e^{-b^k}.$$

As already noted, this mechanism is a major factor in the intermediate temperature range. In the general case, as is seen from equation (1.40), the temperature dependence is very complicated. We will thus consider the main limiting cases.

If, for example, $b^2 \gg b^1$ (more precisely, $\exp(b^2 - b^1) \gg a^2/a^1$), then

$$\frac{\varkappa^s}{\varkappa^n} = \frac{F(b^1)}{F(T_c)},$$

where the function F is determined by equation (1.14). In this case the major factor is the scattering of phonons by the band with the minimal energy gap. Qualitatively this is understandable since such a band would possess the largest number of electron excitations, and therefore the free path of the phonons associated with scattering by the electrons of this band would be the shortest.

When the energy gaps are equal, $b^1 = b^2 = b$ and $I(T) = (a^1 + a^2)^{-1} F(T)$. In this case we naturally arrive at equation (1.14), which describes the usual single-band superconductor.

The lattice thermal conductivity rises exponentially with decreasing temperature below T_c, up to a maximum in the temperature range where $\varkappa_{pe}^s \approx \varkappa_{pd}$ ($\varkappa_{pd}$ is determined by the scattering of phonons by defects, impurities, and the specimen's boundaries). It can be seen from (1.39) and (1.40) that the rise in the multiband case is slower than in the simple case. This fact is associated with the dominant effect of the band with the minimum energy gap.

Introduction of an impurity equalizes the energy gaps of the different bands. Thus, the introduction of an impurity must result in a more abrupt rise in the lattice conductivity with decreasing temperature.

If a superconductor in which $b^1 \ll b^2$ is studied in the temperature range sufficiently far from T_c, so that $b^i \gg 1$ (where $b^i \approx \Delta^i(0)/T$), the ratio $\varkappa^s/\varkappa^n$ is described by the very simple relationship $\varkappa_p^s/\varkappa_p^n \sim \exp(b^1)$. This equation directly relates the minimum energy gap value, $\Delta^1(0)$, to the experimentally measured ratio $\varkappa_p^s/\varkappa_p^n$. Hence measurements of the lattice conductivity can be used together with specific heat and ultrasonic attenuation measurements (see Chapter III) to determine the energy gap $\Delta^i(0)$, corresponding to overlapping energy bands.

10. INVESTIGATION OF THE INTERMEDIATE STATE

A marked decrease in thermal conductivity is experimentally observed during the transition into the intermediate state. This phenomenon was first observed by Mendelssohn and Olsen /45, 65/, who studied specimens of Pb + 0.1 at. %Bi. This effect was also noticed in a number of other experiments, for example in /57, 66–68/, where the decrease in $\varkappa^s$ was observed in single crystals of Hg, Sn, and In. This effect was studied in a wide temperature range in /57, 69, 70/. The decrease in thermal conductivity was noted both at the low temperatures (where the phonon thermal conductivity is the main component), and in the higher temperature range (where the electronic component dominates). In the first case, considered in /71, 72/, the decrease in $\varkappa$ is due to a decrease in the free path of the phonons as they pass from the superconducting into the normal phase. (In these experiments the normal phase regions were perpendicular to the axis of the specimen along which the heat flowed. Consequently, the phonons crossed the interface between the normal and superconducting phases.) The mean free path of the phonons decreased because of the presence of the normal regions, and this led to a decrease in the observed heat flux.

We next consider the higher temperature region, where the electronic component is dominant (the specimen is assumed to be sufficiently pure). In Andreev's experiments /73/ it was shown that specific reflection of electron excitations occurs at the interface between the normal and superconducting phases, across which heat flows. The following equations were investigated:

$$\begin{aligned} i\frac{\partial f}{\partial t} &= -\left(\frac{\nabla^2}{2m}+\mu\right)f + i\Delta(\boldsymbol{r})\varphi, \\ i\frac{\partial \varphi}{\partial t} &= \left(\frac{\nabla^2}{2m}+\mu\right)\varphi - i\Delta(\boldsymbol{r})f, \end{aligned} \tag{1.41}$$

where

$$\begin{aligned} f(\boldsymbol{r}, t) &= \langle \Phi_0 | \psi(\boldsymbol{r}, t) | \Phi_1 \rangle, \\ \varphi(\boldsymbol{r}, t) &= \langle \Phi_0 | \psi^+(\boldsymbol{r}, t) | \Phi_1 \rangle \end{aligned}$$

(ψ^+ and ψ are Heisenberg operators; Φ_0 and Φ_1 are wave functions in the space of occupancy numbers, which describe the ground and excited states of the system respectively; μ is the chemical potential). These equations are obtained from the equations of motion

$$\begin{aligned} &\left(i\frac{\partial}{\partial t}+\frac{\Delta}{2m}+\mu\right)\psi - g(\psi^+\psi)\psi = 0, \\ &\left(i\frac{\partial}{\partial t}-\frac{\Delta}{2m}-\mu\right)\psi^+ + g\psi^+(\psi^+\psi) = 0. \end{aligned}$$

The set of values for f and φ represents the wave function of a quasi-particle.

Analysis of equation (1.41) leads to the conclusion that the momentum hardly changes on reflection, but ξ and the velocity $\partial\varepsilon/\partial\boldsymbol{p}$ reverse their sign. In other words, a transition of an "electron" into a "hole" and vice versa takes place on reflection.

Indeed, the solution of equation (1.41) for the functions f and φ at the relevant quasi-particle energies ($\sim \varepsilon_F$; this enables us to consider the medium as being quasi-homogeneous) can be sought in the form

$$f = e^{ip_F\boldsymbol{nr} - i\varepsilon t}\,\eta(\boldsymbol{r}), \qquad \varphi = e^{ip_F\boldsymbol{nr} - i\varepsilon t}\,\chi(\boldsymbol{r})$$

(p_F is the Fermi momentum). This leads to the following equations for η and χ:

$$\left(ivn\frac{\partial}{\partial r}+\varepsilon\right)\eta-i\Delta(r)\chi=0,$$
$$\left(ivn\frac{\partial}{\partial r}-\varepsilon\right)\chi-i\Delta(r)\eta=0, \tag{1.42}$$

and asymptotic equations for $z\rightarrow\infty$ (when we set $\Delta(r)=0$) which have the form

$$\begin{pmatrix}\eta\\ \chi\end{pmatrix}=A\begin{pmatrix}1\\ 0\end{pmatrix}e^{ik_1r}+B\begin{pmatrix}0\\ 1\end{pmatrix}e^{ik_2r} \tag{1.43}$$

(A and B are arbitrary constants; $k_1=n\varepsilon/v$; $k_2=-n\varepsilon/v$).

From (1.42) and (1.43) it is evident that the first term in equation (1.43) actually describes a "particle" moving in the direction of the unit vector n, and the second term corresponds to a "hole," whose velocity is opposite to n.

The peculiar character of the reflection at the interface between the normal and superconducting phases injects important special features in investigation of the phenomena of interest, heat transfer and sound attenuation (see Chapter III), in the intermediate state. The Andreev reflection was most directly observed by Krylov and Sharvin in the study of microwave size effects /74/. In this work the surface impedance was measured for the n-layer, bounded on one side by a vacuum and on the other side by the s-phase. When the free path is long, the surface impedance is highly sensitive to the nature of the reflection at the interface with the s-phase.

In /75/ the influence of the Andreev reflection on the electric conductivity of the normal layers was investigated. At $T\ll T_c$ the energy of the thermal excitations in the normal phase was less than the energy gap, and they were reflected in the above specific manner with a change in the sign of the charge and the mass, but with the momentum conserved. At $T\approx T_c$, there was no total reflection and the excitations could be transmitted through the s-phase. In /75/ the resulting contribution to the electric conductivity of the normal layers was observed.

We now consider the temperature range $T_0\ll T\ll T_c$ (T_0 is the temperature at which the phonon contribution to the thermal conductivity is not negligible). Here the major factor are the excitations with energy ε close to Δ (the number of the other excitations is exponentially small). In this case the reflection coefficient is

$$w=1-f(n_z)\left(\frac{\varepsilon-\Delta}{\Delta}\right)^{1/2} \tag{1.44}$$

($f \sim 1$, n_z is the unit vector of the normal). The coefficient w is, as we see, close to unity and consequently, heat transfer across the interface between the superconducting and the normal phase is significantly hindered. This situation is similar to heat transfer between a solid body and liquid helium /76, 77/.

Heat exchange between a solid body and liquid He II is hindered because of the great difference between the speeds of sound in the two media. Therefore, phonons pass from He II into the solid and back only to a very small extent (this is even more so for the conversion of a roton into a phonon), since energy and momentum are not conserved.

In the intermediate state the large value of the reflection coefficient w hinders heat transfer. In both cases a temperature jump arises which is related in the intermediate state to the heat flux Q by the following expression:

$$Q = \frac{\sqrt{\pi}}{2} f_0 \left(\frac{p_F}{\pi}\right)^2 \Delta \left(\frac{\Delta}{T}\right)^{1/2} \exp\left(-\frac{\Delta}{T}\right) \delta T.$$

Here $Q = (\partial W/\partial T)\, \delta T$ (W is calculated from

$$W = \int\limits_{v_z > 0} 2n_0(\varepsilon)\, \varepsilon\, (1 - w)\, v_z \frac{dp}{(2\pi)^3},$$

where $n_0(\varepsilon) = e^{-\varepsilon/T}$, and w is determined by (1.44)); $f_0 \sim 1$. The incremental thermal resistivity is determined by the equation

$$\Delta R = f_0^{-1} \left(\frac{2\pi}{p_F}\right)^2 \left(\frac{\varphi(\eta)\, T}{\pi a\, d\Delta^3}\right) \exp\left(\frac{\Delta}{T}\right) \tag{1.45}$$

($a = (\alpha d/\varphi(\eta))^{1/2}$ is the spatial period of structure in the intermediate state /78/; it is assumed that two temperature jumps δT occur in each period a). The temperature dependence of ΔR (in which, as is seen from (1.45), the electronic free path does not appear) fits the experimental data fairly well. In (1.45) d is the diameter of the specimen. The expressions for φ and α are given in /78/.

When $a^s \ll a^n$ (a^s and a^n are the thicknesses of the superconducting and normal layers respectively), the tunneling of the thermal excitations through the s-layer is substantial for $T \ll T_c$. This fact results in a power dependence of ΔR on T.

The specific nature of the reflection of excitations leads to certain peculiar features of the heat transfer along the interface between the superconducting and normal phases /80/. If $T \ll T_c$, the electron component of the normal phase $\varkappa_e^n$ makes the major contribution to the heat flux. The electronic thermal conductivity $\varkappa_e^s$ in this temperature range will be exponentially small (see Secs. 2 and 3). The lattice heat flux $\varkappa_p$ will be larger than $\varkappa_e^s$ in the normal and even more so in the superconducting phases. Nevertheless, its value will amount to only several percent of the electronic heat flux in the normal phase, which is described by $\varkappa_e^n$.

To calculate $\varkappa_e^n$ (in the case of heat transfer along the interface between the normal and superconducting layers) we should start from the kinetic equation for the distribution function $f(\boldsymbol{r}, \boldsymbol{p})$ of the electron and hole excitations in the normal phase:

$$(\partial f/\partial \boldsymbol{r})\, \boldsymbol{v} = -(f-\bar{f})/\tau$$

(the averaging is carried out over a surface of constant energy).

We write the function f in the form

$$f = f_0(T(x_i)) + \chi \partial n_0/\partial T$$

($n_0 = (e^{\varepsilon/T}+1)^{-1}$ is the equilibrium distribution function, $T(x_i)$ is a linear function of the coordinates, whose derivatives with respect to coordinates coincide with the corresponding derivatives of the local temperature; we choose the plane $z = 0$ in the middle of one of the normal layers, so that the x_i run through the values of x and y). Next, setting

$$\chi = -\tau v_i \partial T/\partial x_i + \nu, \tag{1.46}$$

we arrive at the following equation for the function ν:

$$\partial \nu/\partial z + \nu/v_z\tau = 0,$$

whose general solution has the form

$$\nu = C(\boldsymbol{n}, \xi)\, e^{-z/v_z\tau}, \qquad \boldsymbol{n} = \frac{\partial \varepsilon/\partial \boldsymbol{p}}{|\partial \varepsilon/\partial \boldsymbol{p}|}. \tag{1.47}$$

To solve this problem, it is assumed that the electronic free path l is significantly greater than the thickness of the normal layers a^n. In this context we mention an experiment /57/ in which the disappearance of the minimum $\varkappa$ in Sn—In alloys was observed as the In concentration was increased. In addition, the condition $l \ll R$ is assumed (R is the Larmor radius of an electron in the critical magnetic field). This inequality enables us to neglect terms which contain the magnetic field strength in the kinetic equation.

The boundary conditions

$$f(\boldsymbol{n}, \xi) = f(\boldsymbol{n}, -\xi) \text{ for } z = \pm a^n/2.$$

are based on the special nature of the reflections at the interface between the superconducting and normal phases (converting an "electron" into a "hole" and vice versa). Using these boundary conditions we find the function $C(\boldsymbol{n}, \xi)$.

Next, with the aid of (1.46) and (1.47), we find the function $\chi(\boldsymbol{n}, \xi)$, which determines the perturbation of the electron distribution

$$\chi(\boldsymbol{n}, \xi) = -\tau v_i \frac{\partial T}{\partial x_i}\left[1 - \frac{e^{-z/v_z\tau}}{\mathrm{ch}\,(a^n/2v_z\tau)}\right].$$

We next calculate the volume-averaged heat flux density

$$\bar{q}_i = \frac{1}{a}\int_{-a^n/2}^{a^n/2} q_i(z)\,dz,$$

where q_i, as usual, is defined by the equation

$$q_i = 2\int \varepsilon v_i n \frac{d\boldsymbol{p}}{(2\pi)^3}.$$

After substituting the distribution function, the thermal conductivity $\varkappa_{ik}$, is calculated:

$$\varkappa_{ik} = \frac{7\zeta(3)}{12\pi^3} T \frac{(a^n)^2}{a}\int \frac{d(\cos\theta)\,d\varphi}{K(\theta, \varphi)} n_i n_k \delta(\cos\theta). \tag{1.48}$$

Here $K(\theta, \varphi)$ is the Gaussian curvature of the Fermi surface, and the relative orientation of the z-axis and the vector $\boldsymbol{n}$ is determined by the angles θ and φ. The main contribution to the heat flux is clearly from the excitations which propagate parallel to the interface between the superconducting and normal phases. Integrating (1.48) over θ, we find

$$\varkappa_{ik} = \frac{7\zeta(3)}{6\pi^2} T \frac{(a^n)^2}{a} B_{ik}, \tag{1.49}$$

$$B_{ik} = \int_0^{2\pi} \frac{\partial\varphi}{2\pi} \frac{n_i n_k}{K(\varphi)}, \quad K(\varphi) \equiv K\left(\frac{\pi}{2}, \varphi\right) \sim p_F^{-2}.$$

As expected, the electronic thermal conductivity of the normal phase, calculated in the relaxation time approximation, varies linearly with the temperature. Moreover, the intermediate state condition in this case $l \gg a^n$ (as is seen from (1.48) and (1.49), $\varkappa_{ik}$ is not dependent on the free path l in this case) appears in the fact that the heat flux is $la/(a^n)^2$ times smaller than the thermal conductivity of an ordinary normal metal.

It is especially important to consider the boundary region of the intermediate state. In this case the energetically more favorable state is a filamentary (rather than layered) structure of the normal regions.

The calculation of the heat flux along the interface between the normal and superconducting regions is basically carried out by the same method as in the case of a layered structure. However, according to the symmetry of this case, we should switch to a cylindrical coordinate system. As before, the electrons of the normal regions make the main contribution to the heat flux.

The excitations moving at a small angle to the axis of the filament make the major contribution. It is assumed that $l \gg d$ (d is the radius of the filaments). Moreover, the thermal conductivity is dependent on the free path,

$$\varkappa = \eta \frac{T}{24} \frac{a^2 \ln (l/a)}{\bar{K}(0)\, l\,(0)}, \tag{1.50}$$

where η is the concentration of the normal phase; $\bar{K}(0) = \bar{K}(0, \varphi)$, $\bar{K}^{-1}(t, \varphi) = K^{-1}(\arcsin t, \varphi) + K^{-1}(\pi - \arcsin t, \varphi)$. It can be seen that the thermal conductivity decreases with increasing free path length and is proportional to $\ln (l/a)/l$.

11. THERMAL SWITCHES

The specific heat-transfer properties of superconductors can be used in various areas of low-temperature physics. We pause here to describe the so-called thermal switches, which are used primarily to produce low temperatures.

At temperatures near the absolute zero (see Sec. 5), the major factor is the lattice thermal conductivity $\varkappa_{pd}^{s}$, which is caused by the scattering of phonons by impurities and defects and phonon reflection from the crystal boundaries. When the free path is constant (such as when the scattering by the specimen's boundaries is the major factor), $\varkappa_{pd}^{s}$ is proportional to T^3. In a normal metal at the same temperature (the transition into the normal state can be achieved by applying an appropriate magnetic field), the major factor is the electronic thermal conductivity, $\varkappa_{ed}^{n}$, caused by the scattering of electrons by impurities, defects, and boundaries. $\varkappa_{ed}^{n}$ decreases linearly, $(\varkappa_{ed}^{n})^{-1} = \beta/T$. Also $\varkappa^{n}/\varkappa^{s} = \lambda T^{-2}$, where λ is a proportionality coefficient, whose value depends on the kind of substance and can reach values of the order of several hundred. For example, lead wire can be prepared for which $\varkappa^{n}/\varkappa^{s} \approx 10^{5}$ at $T = 0.1°$.

Thus we see that at temperatures near the absolute zero a situation is possible whereby the thermal conductivity in the normal state exceeds the thermal conductivity in the superconducting state at the same temperature by several orders of magnitude. Therefore, converting a substance from the normal to the superconducting state by a suitable change in the external magnetic field can actually make it a thermal insulator. The heat flow is stopped in the circuit of which the substance is a part, since it is now in the normal state.

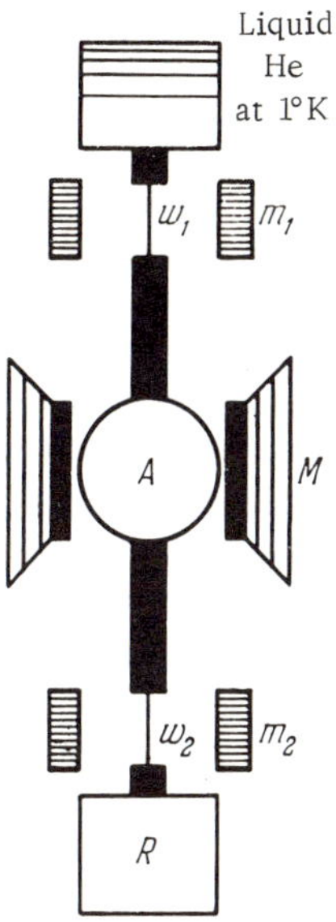

FIGURE 17. Magnetic cooling device /101/:

R) the object being cooled; A) paramagnetic salt; M) magnet; w_1 and w_2) thermal switches activated by magnets m_1 and m_2.

A diagram of a magnetic cooling device developed by Daunt and Heer /101/ is shown in Figure 17. It cools an object by taking advantage of the special thermal conductivity properties of the superconductors used in its thermal switches.

The device functions as follows. Let switch w_1 be in the normal state and switch w_2 in the superconducting state at a given time. Heat can only pass through switch w_1, and the object R is insulated. The method of adiabatic demagnetization is used to change the temperature. Magnet M magnetizes the paramagnetic salt A. The heat evolved is removed to the helium bath through switch w_1. Switch w_1 is then converted into the superconducting state, and switch w_2 is changed into the normal state by magnets m_1 and m_2. Next the paramagnetic salt A is demagnetized, and its temperature decreases. Since switch w_2 is now on, the object R can be cooled by the salt A. The reversal of the switches and the magnetization of the salt are then performed again. Such a periodic device can cool the object R to a temperature limited only by the influx of external heat.

The use of thermal switches in the adiabatic demagnetization method makes it possible to achieve very low temperatures (for further details, see /102/).

Chapter II

THERMAL EFFECTS IN SUPERCONDUCTORS

1. THE THERMOELECTRIC EFFECT

Crystal anisotropy makes it possible to observe a unique thermoelectric effect in superconductors. Formation of a temperature gradient in the specimen gives rise to a normal current $\boldsymbol{j}^n$. In the isotropic case, however, this current is completely canceled by the opposing superconducting current $\boldsymbol{j}^s$, and therefore both the total current and the resulting magnetic field vanish. Thus, a temperature gradient causes only the usual heat conduction and a small additional convection effect (see Sec. 3).

In the anisotropic case, when not one, but several favored directions are associated with grad T, the current $\boldsymbol{j}^n$ need not be completely canceled by the current $\boldsymbol{j}^s$. This results in the appearance of a persistent current and the associated magnetic field, which can be measured experimentally (see below).

The thermoelectric effect was first discussed by Ginzburg /103, 104/, who showed via a suitable generalization of the London theory that consideration of anisotropy invalidates the usual conclusion that thermoelectric phenomena are absent in superconductors.

The thermoelectric coefficients, whose values indicate whether the effect can be observed, can be calculated using the microscopic theory. They cannot be determined from experiments with normal metals since they depend essentially on the parameters of the superconducting state.

In /105/ the thermoelectric effect is discussed on the basis of the modern microscopic theory, and the possibility of its experimental detection is studied. A temperature gradient gives rise to a normal current $\boldsymbol{j}^n_\alpha = b_{\alpha\beta}\ \mathrm{grad}_\beta\ T$ ($b_{\alpha\beta}$ are the thermoelectric coefficients). The basis for solution of the problem is Maxwell's equation and the formulas for the superconducting and normal currents:

$$\mathrm{rot}\, H = \frac{4\pi}{c}(\boldsymbol{j}^n + \boldsymbol{j}^s), \tag{2.1}$$

$$j_\alpha^n = b_{\alpha\beta}\,\mathrm{grad}_\beta T, \tag{2.2}$$

$$j_\alpha^s = -\frac{c}{4\pi}K_{\alpha\beta}(q)\,A_\beta(q) \tag{2.3}$$

($K_{\alpha\beta}$ is the Pippard tensor, relating the current to a vector potential; $K_{\alpha\beta}$ is calculated using the microscopic theory /106/).

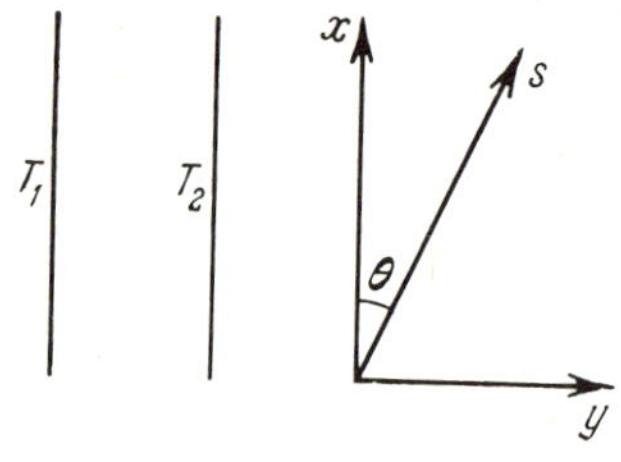

FIGURE 18. Superconducting sheet (s is the crystal axis).

Consider an infinite superconducting sheet (Figure 18). It is clear by symmetry that the components j_z^n and j_z^s, as well as the corresponding quantities $b_{xz}, b_{zy}, K_{xz}, K_{zy}$, etc., vanish. From the continuity equation it follows that in the plane case the component $j_y = j_y^n + j_y^s$ of the total current also vanishes. The only nonzero component is j_x; by (2.1) and (2.3) this component is given in the half-space $y > 0$ by

$$j_x(y) = (cH/4\pi\delta)\,e^{-y/\delta} \tag{2.4}$$

(H is the magnetic field across the thickness of the specimen (see below), $\delta = [K_{xx}(0)]^{-1/2}$).

The relationship obtained in the plane case is naturally gauge-invariant. For a sheet $-a \leqslant y \leqslant a$ (see Figure 18) we find

$$j_x = (cH/4\pi\delta)\ \mathrm{sh}\ (y/\delta)/\mathrm{ch}\ (a/\delta) \tag{2.5}$$

in agreement with /103/. Thus, the anisotropy of the crystal gives rise to a nonzero persistent current in a surface layer of thickness $\sim \delta$.

We next calculate the magnetic field induced by the thermoelectric current. From (2.1)–(2.3) we find

$$A_x(y) = \frac{4\pi}{c}\int \frac{j^n(q)\,e^{iqy}dq}{q^2 + K(q)}, \tag{2.6}$$

where

$$K(q) = K_{xx}(q) - K_{xy}^2(q)/K_{yy}(q), \tag{2.7}$$

$$j^n(q) = j_x^n(q) - (K_{xy}(q)/K_{yy}(q))\,j_y^n(q). \tag{2.8}$$

It is significant that the components of the Pippard tensor corresponding to $q \to 0$ appear in the magnetic field formula (2.6) even for Pippard superconductors, i. e., in the study of thermoelectricity all superconductors are London. The reason for this is that in the possible range of the temperature gradient (grad $T \lesssim 0.1$ deg/cm) the normal current associated with grad T varies only slightly over distances of the order of ξ_0. Thus, the dominating role in the equation

$$A_x(y) \sim \int j^n(\boldsymbol{q}) e^{iqy} [q^2 + K(q)]^{-1} d\boldsymbol{q},$$

which determines the vector potential, is played by values of q much less than ξ_0^{-1}. In addition, of course, $q \ll \delta^{-1}$, except for a very small temperature interval near T_c.

The magnetic field is thus described by the following relationship:

$$H_z(y) = -\frac{4\pi}{c}\frac{d}{dy}(K^{-1}(0, T))\, j^n(y). \tag{2.9}$$

In the isotropic case the field H_z, as expected, vanishes. Indeed, in this case the values of j_x^n and K_{xy}, which determine the value of j^n according to equation (2.8), are equal to zero.

Thus, the problem is reduced to calculation of the normal current or, in other words, to determination of the thermoelectric coefficients $b_{\alpha\beta}$ which characterize the anisotropic superconductor. The components of the normal current can be written

$$j_x^n = (b_\parallel - b_\perp)\cos\theta\sin\theta\nabla T,$$

$$j_y^n = (b_\parallel \sin^2\theta + b_\perp \cos^2\theta)\nabla T.$$

Here $b_\parallel$ and $b_\perp$ are the principal values of the tensor $b_{\alpha\beta}$; $b_\parallel$ corresponds to $\nabla T \parallel s$.

By definition

$$j^n = -e\int \frac{\partial \varepsilon}{\partial \boldsymbol{p}} f^{(1)} \frac{d\boldsymbol{p}}{(2\pi)^3}, \tag{2.10}$$

where $f^{(1)}$ is an increment to the electronic distribution function, whose form is determined by solving the kinetic equation (see Chapter I, Sec. 2)

$$-\frac{\varepsilon}{T}\frac{\partial f^{(0)}}{\partial \varepsilon}\frac{\partial \varepsilon}{\partial \boldsymbol{p}}\nabla T = I_{\mathrm{col}}, \quad f^{(0)} = (e^{(\varepsilon-\mu)/T} + 1)^{-1}.$$

The terms representing the effect of the magnetic field on the distribution function are here proportional to $(\nabla T)^3$ and may be neglected.

We now consider the scattering of electron excitations by impurities. Since the factor disturbing the electronic system is a temperature gradient, the problem is similar to calculation of the electronic thermal conductivity of a superconductor, which is discussed in the isotropic case in /14, 16/. Note that a discussion based on the equation for the thermoelectric coefficient /107/

$$b = (2\omega T)^{-1} \int d^4x \langle [\hat{j}(x), \hat{q}(0)] \rangle,$$

where $\hat{j}$ and $\hat{q}$ are the operators of current density and heat flux, leads, as does calculation of the thermal conductivity /17/, to the same result as the kinetic equation method.

Applying the Bogolyubov (u, v)-transformation, we find the following expression for the Hamiltonian of the interaction of electron excitations of a superconductor with the impurity atoms:

$$H' = \sum_{p,p'} (u_p u_{p'} - v_p v_{p'}) V_{pp'} (\overset{+}{\alpha}_{p'0} \alpha_{p0} + \overset{+}{\alpha}_{p'1} \alpha_{p1}),$$

$$u_p^2 = {}^1\!/_2 (1 + \xi_p/\varepsilon_p), \qquad v_p^2 = {}^1\!/_2 (1 - \xi_p/\varepsilon_p).$$

Allowing for the elastic nature of the scattering $(\varepsilon_p = \varepsilon_p)$ and assuming in addition that $\Delta(p) \approx \Delta = \text{const}$, which is permissible since the energy gap anisotropy is smaller than the anisotropy of the Fermi surface, we find (see also /60/)

$$-\frac{\varepsilon}{T} \frac{\partial f_0}{\partial \varepsilon} \frac{\partial \varepsilon}{\partial p} \nabla T = -2\pi \frac{|\xi|}{\varepsilon} \int |V|^2 (f_p^{(1)} - f_{p'}^{(1)}) \frac{d\sigma}{v_F}, \tag{2.11}$$

where $d\sigma$ is an element of the Fermi surface /106/.

The solution of (2.11) can be written (see, for example, /108/)

$$f^{(1)} = -\frac{\partial f^{(0)}}{\partial \varepsilon} \frac{\varepsilon}{T} (\Lambda \nabla T), \tag{2.12}$$

where $\boldsymbol{\Lambda} = \tau(\boldsymbol{p})\boldsymbol{v}$ is the free path vector length, τ the relaxation time, and $\boldsymbol{v}$ the Fermi velocity of normal electrons. (The electronic free paths in the normal and superconducting states coincide.)

We next calculate the normal current, using (2.12). Integrating over $d\varepsilon$, we find

$$j^n = \frac{2e}{T} \mathcal{F}\left(\frac{\Delta}{T}\right) \int \frac{d\sigma}{v_F} \boldsymbol{v} (\boldsymbol{\Lambda} \nabla T), \tag{2.13}$$

where

$$\mathcal{F}(x) = x(e^x + 1)^{-1} - \ln(1 + e^{-x}). \qquad (2.14).$$

We determine the thermoelectric coefficients $b_\parallel$ and $b_\perp$ from (2.13):

$$b_{\parallel,\perp} = \frac{2ek}{(2\pi)^3} \mathcal{F}\left(\frac{\Delta}{T}\right) \int \frac{d\sigma}{v_F} v^2_{\parallel,\perp} \tau(\boldsymbol{n}). \qquad (2.15).$$

Thus, the thermoelectric coefficients depend on the superconducting state parameters.

From (2.9), (2.7), and (2.8), we obtain

$$H_z(y) = -\frac{4\pi}{c} (\nabla T)^2 \frac{dR}{dT}, \qquad (2.16)$$

where

$$K(0) R = (b_\parallel - b_\perp) \cos\theta \sin\theta - \frac{K_{xy}}{K_{yy}} (b_\parallel \sin^2\theta + b_\perp \cos^2\theta),$$

and $b_{\parallel,\perp}$ are determined from (2.15). Expressing the components of the Pippard tensor which appear in this equation in terms of the principal values, we arrive at the following final equation:

$$H_z(y) = \frac{2\pi kc}{e} (\nabla T)^2 (\varphi_\parallel - \varphi_\perp) \frac{d}{dT} \left[\mathcal{F}\left(\frac{\Delta}{T}\right) \frac{\delta_L^2(T)}{\delta_L^2(0)} \right] \sin^2\theta,$$

$$\varphi_{\parallel,\perp} = \int \frac{d\sigma}{v_F} v^2_{\parallel,\perp} \tau(\boldsymbol{n}) \Big/ \int \frac{d\sigma}{v_F} v^2_{\parallel,\perp}. \qquad (2.17)$$

It is significant that the thermoelectric magnetic field is expressed in terms of the universal function T/T_c and the quantities that characterize a normal metal. In the temperature range near the critical point, equation (2.17) becomes

$$H = -\pi \frac{ck}{e} (\nabla T)^2 (\varphi_\parallel - \varphi_\perp) \frac{1.4}{T_c} \sin 2\theta \left(1 - \frac{T}{T_c}\right)^{-2}. \qquad (2.18).$$

It is evident that the effect increases as $T \to T_c$, except for a very small interval ΔT near T_c where the persistent current disappears (see above). One sees from (2.17) or (2.9) that the effect disappears with transition to the isotropic case.

Let us estimate the possible value of the magnetic field:

$$H \sim \frac{ck}{e} (\nabla T)^2 \frac{\Lambda}{v_F T_c} \left(1 - \frac{T}{T_c}\right)^{-2}.$$

If, say, $\nabla T \approx 0.1$ deg/cm and $\Lambda \approx 0.1$ cm (see, for example, /15/; the specimen must consequently be fairly pure), then when $T/T_c \approx 0.99$ the possible fields are $H \sim 10^{-3}$ gauss, and when $T/T_c \approx 0.9$ the fields are $H \sim 10^{-5}$ gauss.

With decreasing temperature or upon doping of the specimen the thermoelectric magnetic field and the magnetic moment decrease simultaneously. In light of this, the thermoelectric effect in superconductors yields a simple method for experimental detection of the quantization of the magnetic flux /110/.

The strongest thermoelectric effect appears when the following conditions are fulfilled: 1) the specimen must be sufficiently pure; 2) the crystal should be uniaxial; 3) the temperature must be near T_c.

The values given above for the magnetic field H can be detected experimentally, so that it should be interesting to set up appropriate experiments.

2. THE THERMOMAGNETIC EFFECT

Thermomagnetic effects arise from the action of a magnetic field on the heat flux (see, for example, /111/). The Leduc-Righi effect, involving the appearance of a temperature gradient perpendicular to the direction of the resultant heat flux, may be observed in superconductors.

To study this effect /112/, we write the kinetic equation for electron excitations in a temperature gradient along the x-axis and perpendicular to the heat flux caused by a strong magnetic field:

$$-\frac{\partial f}{\partial \varepsilon}\frac{\varepsilon}{T}v_x\frac{\partial T}{\partial x}+\frac{eH}{c}\left(v_y\frac{\partial f}{\partial p_x}-v_x\frac{\partial f}{\partial p_y}\right)\frac{\xi}{|\xi|}=-\frac{f-f_0}{\tau}. \qquad (2.19).$$

The relaxation time τ equals $\tau=\tau_0\,\varepsilon/|\xi|$ (see Chapter I, Sec. 2); it is natural to assume that the dimensions of the specimen are less than the magnetic field penetration depth. Solving equation (2.19) by successive approximations $f=f_0+f_1+f_2$, we find

$$f_1=\frac{p_x}{m}\tau_0\frac{\partial f_0}{\partial \varepsilon}\frac{\varepsilon}{T}\frac{\partial T}{\partial x}\frac{\xi}{|\xi|},\quad f_2=\tau_0^2 v_y\frac{1}{T}\frac{eH}{mc}\frac{\partial f_0}{\partial \varepsilon}\frac{\partial T}{\partial x}\frac{\varepsilon^2}{|\xi|}. \qquad (2.20).$$

The magnitude of the Leduc-Righi effect is determined by the coefficient $L=(\partial T/\partial y')/(\partial T/\partial x')\,H$ (the x'-axis coincides with the direction of the resultant heat flow). It is easy to see that $L=Q_y/Q_xH$, where

$$Q_x=2\int \varepsilon v_x f\,\frac{dp}{(2\pi)^3},\quad Q_y=2\int \varepsilon v_y f\,\frac{dp}{(2\pi)^3}.$$

Using (2.20), we find

$$L = \tau_0 e/mc. \tag{2.21}$$

This ratio is independent of Δ. Consequently, the thermomagnetic Leduc-Righi effect is theoretically observable in superconductors, the corresponding thermomagnetic coefficient being $L = \tau_0 e/mc$.

The Nernst-Ettingshausen effect, consisting in the appearance of an electric field perpendicular to the direction of the resultant heat flux, is clearly absent in superconductors of the first kind (see Chapter III, Sec. 14).

3. CONVECTION

The presence of a temperature gradient results in a small convection effect (see below). The nature of this effect, pointed out in /104, 113, 114/, is as follows. When a temperature gradient ∇T is imposed on the specimen, the usual heat flux (see Chapter I) is accompanied by a normal current (see Sec. 1 of this chapter) created by the normal excitations that participate in the energy transfer determining the thermal conductivity of the superconductor. In the isotropic case, however, this current is completely canceled by the superconducting current, so that $\boldsymbol{j}^n + \boldsymbol{j}^s = 0$.

The motion of the normal and superconducting components is exactly like the circulation of the normal and superfluid components of unevenly heated He II, but the motion of the normal component involves an entropy flow, whereas the superconducting component possesses no entropy. Therefore, although the total charge transfer is indeed zero, the convection mechanism gives rise to a certain finite conductivity.

Let us estimate the value of the convection coefficient /14/. The value of the heat flux by convection Q_c is $Q_c = TSv^n$, or $Q_c = TSj^n_m/\rho^n$ (S is the unit volume entropy, j^n_m the mass flow density due to the normal excitations, and ρ^n the density of the normal component of the superconductor). The entropy of a normal excitation gas is described by the usual combinatorial expression, which can be put in the form

$$S = \frac{2\Delta^2}{T} \nu \int \frac{\operatorname{ch} 2\varphi}{e^{\Delta \operatorname{ch} \varphi / T} + 1} d\varphi.$$

Near the critical point, the entropy is given by (1.22'). The mass flow density is calculated by the equation

$$j^n_m = 2m \int v_x f^{(1)} \frac{d\boldsymbol{p}}{(2\pi)^3},$$

where $f^{(1)}$ is the distribution function perturbed by the temperature gradient, found by solving the kinetic equation (1.2) (see Chapter I, Sec. 2). The density ρ^n, like the proportionality factor relating the total momentum of the electron excitations and their macroscopic velocity, is /6/

$$\frac{N^n}{N_0} = 2\frac{\Delta}{T}\int_0^\infty \frac{\operatorname{ch}\varphi\, d\varphi}{(e^{\Delta \operatorname{ch}\varphi/T}+1)(e^{-\Delta \operatorname{ch}\varphi/T}+1)},$$

and in the temperature range near the critical temperature,

$$(N^n/N_0)_{T\to T_c} = 2(1 - T/T_c)$$

(N_0 is the total number of ground state free electrons).

Using the values of S, j^n, and ρ^n thus determined, one can estimate the magnitude of the convection coefficient. It is clear that since the heat flux is transferred by the normal component of the electron fluid, the flux is maximum in the temperature range near the critical point, which involves an exponential decrease in the number of normal excitations with decreasing temperature. Estimating Q_c, in this temperature range, we conclude that Q_c/Q (where Q is the usual heat flow, discussed in Chapter I and defined as

$$Q = \int \varepsilon v_x f^{(1)} \frac{dp}{(2\pi)^3})$$

is of the order $Q_c/Q \sim T_c\nu/N$ (ν is the density of states), or

$$Q_c/Q \sim T_c/\varepsilon_F.$$

Thus, we see that the ratio of the convection coefficient to the thermal conductivity is a quantity of order from 10^{-5} to 10^{-4} even when $T \approx T_c$. Consequently, the convection coefficient is negligibly small compared to the thermal conductivity.

To summarize, the main effect of a temperature gradient is normal heat conduction. In addition, the thermal effects described in this chapter may also be observed.

Chapter III

SOUND ABSORPTION IN SUPERCONDUCTORS

The attenuation of longitudinal sound waves in superconductors had been studied before the advent of superconductor theory (see, for example, Bömmel /115/ and Mackinnon /116/). A study of this effect in Pb and Sn led to the conclusion that the attenuation decreases monotonically when the temperature falls below T_c. According to the two-fluid model of Gorter and Casimir the attenuation coefficient γ^s is proportional to $(T/T_c)^4$. However, the experimentally observed function $\gamma^s(T)$ is significantly more abrupt.

Sound attenuation was one of the first questions tackled by the superconductor theory, and this was not accidental. The point is that the corresponding measurements furnish a wealth of information about the specific properties of the electron excitation spectrum of a superconductor. The dependence of the energy gap on temperature and angle is reliably determined by ultrasonic measurements.

Any discussion of sound attenuation should distinguish two limiting cases. The first corresponds to frequencies ω such that $\omega \gg \tau^{-1}$, where τ is the relaxation time, and the second to $\omega \ll \tau^{-1}$ (long wavelength sound). In the first case relaxation processes may generally be neglected, and sound absorption can be viewed simply as a process of absorption of sound quanta by free electron excitations. The attenuation is then completely analogous to Landau's well-known plasma wave attenuation /117/ and actually represents the absorption of waves by resonance electrons. In the second case, of course, the relaxation processes play a decisive role, and the energy dissipation of a sound wave is determined by investigating the appropriate kinetic equation.

1. ABSORPTION OF HIGH-FREQUENCY ULTRASOUND

In this case the sound wave period is significantly smaller than the relaxation time. As noted, the interaction of a wave with the electronic system can be viewed as radiation and absorption of sound quanta by electron excitations.

We set out from the electron-phonon interaction Hamiltonian, transformed to the new Fermi operators (see equation 1.7)). We must also remember that the number N of sound quanta is much greater than unity.

Writing the probability of the absorption of a sound quantum with the aid of the (u, v)-transformation (1.3):

$$w_{\rm I} \sim \sum_k |V|^2 [(u_k u_{k'} - v_k v_{k'})^2 N f(1-f') \times \\ \times \delta(\varepsilon' - \varepsilon - \hbar\omega) + (u_k v_{k'} + u_{k'} v_k)^2 N f f' \delta(\varepsilon + \varepsilon' - \hbar\omega)],$$

and the probability of the reverse process:

$$w_{\rm II} \sim \sum_k |V|^2 [(u_k u_{k'} - v_k v_{k'})^2 (N+1)(1-f) f' \times \\ \times \delta(\varepsilon' - \varepsilon - h\omega) + (u_k v_{k'} + u_{k'} v_k)^2 (N+1) f f' \delta(\varepsilon' + \varepsilon - \hbar\omega)],$$

we obtain the following expression for the absorption coefficient /20,32/:

$$\gamma \sim \int |V|^2 \left\{ \left(1 + \frac{\xi\xi' - \Delta^2}{\varepsilon\varepsilon'}\right)(f - f')\,\delta(\varepsilon' - \varepsilon - \hbar\omega) + \\ + \left(1 - \frac{\xi\xi' - \Delta^2}{\varepsilon\varepsilon'}\right)(1 - f - f')\,\delta(\varepsilon' + \varepsilon - \hbar\omega)\right\} \frac{p^2\,dp \sin\vartheta\,d\vartheta\,d\varphi}{4\pi^2\hbar^4}. \quad (3.1)$$

Integrating over the angle and the excitation energy (the approximation (1.13) is also used in the calculation), we arrive at the following equation for the ratio of ultrasonic absorption coefficients in the normal and superconducting states ($\hbar\omega \lessgtr T$):

$$\gamma^s/\gamma^n = \{x - \ln[(e^{b+x}+1)(e^b+1)^{-1}] + D(x)(2b - x + \\ + 2\ln[(e^{x-b}+1)(e^b+1)^{-1}])\}\{\ln[(e^x+1)/2]\}^{-1}, \quad (3.2)$$

where $x = \hbar\omega/T$; $b = \Delta/T$; and the function $D(x)$ (see Chapter I, Sec. 4 above) is introduced because the creation of an excitation pair is possible only when $\hbar\omega \geqslant 2\Delta$. The second term in equation (3.1) and the corresponding term in (3.2), which contains $D(x)$, describe the creation of an excitation pair by a sound quantum (the threshold effect).

When $x = \hbar\omega/T \ll 1$, we obtain (without allowance for the threshold effect) the earlier equation of Bardeen, Cooper, and Schrieffer /1/:

$$\gamma^s/\gamma^n = 2/(e^b+1). \quad (3.3)$$

Equations (3.2) and (3.3) also describe the relaxation of the nuclear quadrupole moment /1/.

The abrupt decrease in absorption, observed experimentally when the temperature falls below T_c, is described, as can be seen from (3.3), by a "Fermi" function. This effect is associated with a rapid increase in the function $\Delta(T)$ as the temperature leaves the critical region, and with a decrease in the number of electron excitations (Figure 19).

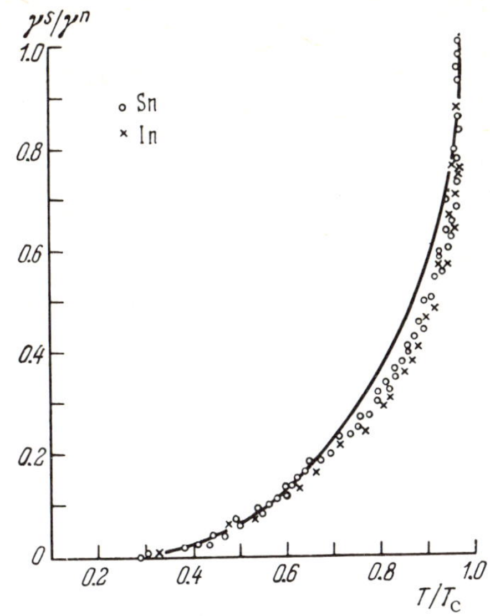

FIGURE 19. Temperature dependence of γ^s/γ^n. The experimental data refer to Sn and In.

If $\hbar\omega \gg T$ the absorption as $T \to 0$ is described by the following asymptotic equation:

$$\gamma^s/\gamma^n = e^{-\Delta/T}\,[2\pi T\Delta/\omega\,(\omega + 2\Delta)]^{1/2},$$

obtained in /118/ by directly calculating the first term on the right of (3.1). It is also assumed that $\hbar\omega < 2\Delta\,(T)$, so that one can disregard the corresponding threshold effects described by the second term in (3.1).

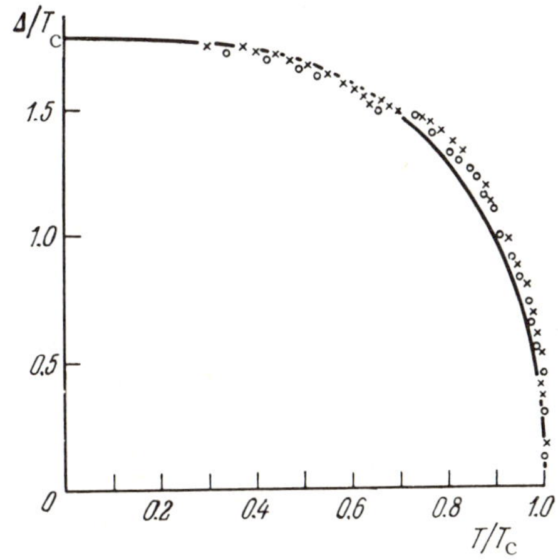

FIGURE 20. Energy gap as a function of temperature. The experimental data refer to ultrasonic attenuation in tin /119/.

Equations (3.2) and (3.3) directly relate the energy gap to the experimentally measured ratio γ^s/γ^n. The simplicity of equation (3.3) makes for highly effective use of ultrasonic measurements to determine the form of the function $\Delta(T)$. Figure 20 shows experimental data /119–121/ obtained in a study of ultrasonic attenuation in Sn single crystals. The excellent agreement between theory and experimental data should be noted. This example clearly demonstrates the advantage of ultrasonic measurements for a more accurate reconstruction of the function $\Delta(T)$, which is a basic characteristic of the superconducting state.

2. EFFECT OF AN ENERGY GAP ANISOTROPY ON ULTRASONIC ATTENUATION

The absorption of ultrasound is highly sensitive to the structure of the energy spectrum of the substance being studied. Therefore, as mentioned above, the corresponding investigations yield quite complete information concerning the dependence of $\Delta\,(T,\,\boldsymbol{n})$ on temperature and angle in the excitation spectrum.

A whole series of experiments bear witness to a marked anisotropy of ultrasonic absorption. Equations (3.2) and (3.3), which were derived in the isotropic approximation, are not applicable to analysis of these data. References /122–124/ are devoted to a theoretical consideration of the effect of anisotropy on ultrasonic attenuation.

2.1. Energy Gap Anisotropy

In the usual superconductivity theory, the electron-phonon interaction matrix element is considered constant, its value being independent of the direction. The electrons obey a square dispersion law. The theory describes the main experimental facts moderately well, but in some cases the idealized model, which involves certain assumptions, turns out to be inadequate.

The energy gap anisotropy is observed directly in experiments by ultrasonic absorption. Certain effects (for example, anisotropy of the penetration depth of a magnetic field in crystals /125, 11, 126/ or the thermoelectric effect, see Chapter II, Sec. 1) are completely determined by the anisotropy and disappear upon transition to the isotropic model. When comparing theory with experiment, it is often necessary to consider the effect of anisotropy, so that discussion of the anisotropic model of a superconductor is of some interest. The effects associated with anisotropy were considered in a general form by Pokrovskii /95/ and independently, within the specific limits of the Fermi surface, in /127, 128/.

The equation for the direction-dependent energy gap $\Delta(\boldsymbol{p})$ is

$$\Delta(\boldsymbol{p}) = g\int u(\boldsymbol{p}, \boldsymbol{p}')\frac{\Delta(\boldsymbol{p}')}{\varepsilon(\boldsymbol{p}')}\,d\boldsymbol{p}',\ \hbar=1, \tag{3.4}$$

and this is the basic equation of the theory. Integrating over ξ' in (3.4), we find /95/

$$\Delta(\boldsymbol{n}) = g\int u(\boldsymbol{n}, \boldsymbol{n}')\,\Delta(\boldsymbol{n}')\ln\frac{2\omega_D}{\Delta(\boldsymbol{n}')}\frac{d\sigma}{v_F'}, \tag{3.5}$$

where $\boldsymbol{n}=\boldsymbol{p}/|\boldsymbol{p}|$, and $d\sigma$ is an area element on the Fermi surface. At finite temperatures the energy gap, which depends on temperature and direction, satisfies the equation

$$\Delta(\boldsymbol{n}) = g\int u(\boldsymbol{n}, \boldsymbol{n}')\left[\ln\frac{2\omega_D}{\Delta(\boldsymbol{n}')} - F(\Delta(\boldsymbol{n}')/T)\right]\Delta(\boldsymbol{n}')\frac{d\sigma'}{v_F'}, \tag{3.6}$$

$$F(x) = \int_{-\infty}^{\infty} dy\,(y^2+1)^{-1/2} f(x\sqrt{y^2+1}), \qquad f(x) = (e^x+1)^{-1}.$$

Equations (3.5) and (3.6) were solved by Pokrovskii in /95/ in the weak coupling approximation. They are linearized by assuming the function $\Delta(\boldsymbol{n})$ to have the form

$$\Delta(\boldsymbol{n}) = 2\omega_D\varphi(\boldsymbol{n})\,e^{-\Lambda/g}.$$

The result is the following general formula for $\Delta(\boldsymbol{n})$:

$$\Delta(\boldsymbol{n}) = 2\omega_D e^{-\Lambda/g}\,Q\psi(\boldsymbol{n}), \tag{3.7}$$

where $\psi(\boldsymbol{n})$ is the normalized solution of the homogeneous integral equation

$$\psi(\boldsymbol{n}) = \Lambda\int u(\boldsymbol{n}, \boldsymbol{n}')\,\psi(\boldsymbol{n}')\frac{d\sigma'}{v_F'}$$

(the normalization condition

$$\int \psi^2(\boldsymbol{n})\,d\sigma/(p_0 v_F) = 1,\ p_0 \sim p_F$$

stems from the equality $\int d\sigma/(p_0 v_F) = 1$, and Λ is the smallest positive eigenvalue of the homogeneous equation).

The quantity Q, determined by calculating the next approximation to (3.7), turns out at $T = 0$ to be

$$Q = \exp\left\{-\int \psi^2(\boldsymbol{n}) \ln \psi(\boldsymbol{n}) \frac{d\sigma}{p_0 v_F}\right\}. \tag{3.8}$$

At finite temperatures the value of Q is temperature dependent. Thus, with a temperature change, the energy gap changes proportionally in various directions. It vanishes simultaneously in all directions at the temperature

$$T_c = 2(\gamma/\pi)\,\omega_D e^{-\Lambda/g}. \tag{3.9}$$

An interesting relationship was obtained in /95/ for the specific heat jump during the transition from superconducting to normal state; in the anisotropic case in the weak coupling approximation, it becomes

$$\beta = \left(\frac{\Delta c}{c^n}\right)_{T_c} = \frac{1{,}4}{B}, \quad B = \int \psi^4(\boldsymbol{n}) \frac{d\sigma}{p_0 v_F} > 1.$$

Consequently, $\beta < \beta_{BCS} = 1.4$. This relationship may be violated only if there is a strong electron-phonon interaction (see Chapter I, Sec. 9). In /95/ several other general relationships describing the anisotropic case were obtained.

The values of Δ_{max} and Δ_{min} differ (see, for example, /129/) by a factor of at most 1.5. Thus, even for a highly anisotropic Fermi surface the anisotropy of the energy gap is not too great. The reason is that the energy gap is an averaged integral characteristic of the superconductor.

The energy gap anisotropy is also found to be small when the matrix element for the electron-phonon interaction is considered. Using symmetry considerations, one can determine the form of the different terms in the expression for the matrix element.

It turns out that the direction-dependent energy gap can be written as $\Delta(\Omega) = \Delta^{(0)} + \Delta^{(1)}(\Omega)$, where $\Delta^{(0)}$ is independent of the angle, and $\Delta^{(1)}(\Omega)$ contains as factors small parameters $a \ll 1$ and $b \ll 1$, which lead to the inequality $\Delta^{(1)} \ll \Delta^{(0)}$. If, for example, a and b are estimated via the strong coupling approximation of the electronic theory of metals, a measure of the relative smallness of the angle-dependent component is the ratio r_0^2/d^2, where r_0 is the Bohr radius and d the lattice constant.

In /127, 128/ the limiting cases of small and large anisotropy are discussed in greater detail. Among these are, first, a simplified model of a closed surface — an ellipsoid of revolution with small anisotropy and, second, a cylindrical Fermi surface, which is a simplified model of an open surface. Explicit expressions are obtained for the angular and temperature dependence of the energy gap in these cases, enabling one, in particular, to consider the behavior of several thermodynamic quantities. We will not dwell on the corresponding equations here.

It is significant that anisotropic effects are most marked in moderately pure superconductors. Introduction of an impurity results in a gradual "isotropization" of the energy gap /30, 130/. During elastic scattering of electrons by impurity atoms, the electrons can pass from one section of the Fermi surface to another, resulting in a further averaging. The anisotropic effects are substantial as long as Anderson's inequality $l \gg \xi_0$ /30/ remains valid (l is the free path and ξ_0 the size of a Cooper pair). When the free path is short the superconductor is adequately described by the isotropic model.

We now proceed directly to our main topic — ultrasonic attenuation in anisotropic superconductors.

2.2. Absorption Coefficient. Construction of the Function $\Delta(n)$

We first confine ourselves, following Pokrovskii /122/, to a pure superconductor. Attenuation is described by the imaginary part of the polarization operator $\Pi(q, i\omega_n)$,

$$\Pi(q,\ i\omega_n) = \frac{T}{(2\pi)^3}\sum_{\omega_n} dp'\,[G(p)G(p-q) - F(p)F(p-q)], \quad (3.10)$$

where $\omega_n = (2n+1)\pi T$. Solution of Dyson's equations leads to the following expressions for the Green's functions in the anisotropic case:

$$G(p,\ \omega_n) = \left(\frac{u_p^2}{\varepsilon_p - \omega_n} - \frac{v_p^2}{\varepsilon_p + \omega_n}\right) g(p),$$

$$F(p,\ \omega_n) = \frac{\Delta(p)}{2\varepsilon_p}\left(\frac{1}{\varepsilon_p - \omega_n} + \frac{1}{\varepsilon_p + \omega_n}\right) g(p)$$

($u_p^2,\ v_p^2 = {}^1/_2\,(1 \pm \xi_p/\varepsilon_p)$; the dimensionless function $g(p)$ depends essentially on the direction).

We pass from summation over ω_n to integration, via the substitution

$$T\sum_{\omega_n} = \frac{i}{4\pi}\int_\Gamma \mathrm{th}\left(\frac{\beta\omega}{2}\right) d\omega$$

(Γ is a contour which consists of two straight lines parallel to the imaginary axis). This gives the following expression for $\mathrm{Im}\,\Pi$ (when $\omega < 2\Delta$):

$$\mathrm{Im}\,\Pi = \frac{1}{(2\pi)^2}\int dp g^2(p)\frac{\xi^2}{\varepsilon^2}[f(\beta\varepsilon-\beta\omega)-f(\beta\varepsilon)]\,\delta(\varepsilon-\varepsilon'-\omega),\ \ \beta = T^{-1}.$$

The required ratio γ^s/γ^n is given by the equation

$$\gamma^s/\gamma^n = 2\langle f(\beta\Delta)\rangle_n \tag{3.11}$$

$$\langle\varphi\rangle_n = \frac{\int \frac{d\sigma'}{v'_F} g^2(n')\,\delta(\cos x)\,\varphi(n')}{\int \frac{d\sigma'}{v'_F} g^2(n')\,\delta(\cos x)}$$

($d\sigma$ is an area element on the Fermi surface, and x the angle between the direction of the Fermi velocity and the direction of the sound propagation), from which it is evident that sound absorption is indeed significantly anisotropic.

A very convenient equation for analysis of the experimental data is obtained by consideration of ultrasonic absorption in the low temperature range, $1 < T_c/T < (2/\alpha)\ln(v_F/u)$ (u is the velocity of sound, α the anisotropy coefficient, which equals the ratio of the change in $\Delta(\boldsymbol{n})$ on the Fermi surface to the minimum Δ_0). This formula is

$$\gamma^s(\boldsymbol{q}) \sim \omega(\boldsymbol{q})\exp\left(-\frac{\Delta\,\mathrm{min}}{T}\right), \tag{3.12}$$

where $\omega(\boldsymbol{q})$ is the phonon frequency, and $\Delta_{\min}(\boldsymbol{n})$ the minimum energy gap on the circumference of the stereographic projection of the Fermi surface perpendicular to the vector $\boldsymbol{n} = \boldsymbol{q}/q$.

At lower temperatures, where measurements have apparently not yet been carried out, sound absorption is determined by the absolute minimum energy gap Δ_0:

$$\gamma \sim \exp(-\beta\Delta_0)\sum|\cos^3\chi_0|^{-1}$$

(χ_0 are the angles between the velocities at the absolute minimum point and the direction of $\boldsymbol{q}$).

One sees directly from equation (3.12) that the absorption coefficient γ^s decreases exponentially with the temperature, the exponent depending on the direction. According to this equation the absorption depends on the minimum of the energy gap along the line $\boldsymbol{qv} = 0$. This result has an obvious physical interpretation:

in the absorption of high-frequency sound, which may be viewed (see Sec. 1) as a straightforward quantum process of phonon absorption by electrons, energy must be conserved:

$$\varepsilon_{p+q} - \varepsilon_p = \omega_q, \quad \omega_q = uq.$$

Recalling that $u \ll v_F$, we can write this condition in the form

$$\boldsymbol{vq} = \omega. \tag{3.13}$$

As expected, electrons moving in the equal-phase plane of the sound wave thus make the major contribution to resonance absorption. Since the velocity of sound satisfies the inequality $u \ll v_F$, condition (3.13) can be written approximately as $\boldsymbol{vq} = 0$, i.e., electron excitations with velocities perpendicular to the direction of the sound propagation indeed make the major contribution to absorption. At low temperatures, the Boltzmann distribution of superconductor electron excitations leads to equation (3.12), which involves the minimum of the energy gap along the line $\boldsymbol{vq} = 0$.

Pokrovskii and Toponogov /123/ discussed the possibility of reconstructing the energy gap $\Delta(\boldsymbol{n})$ from ultrasonic attenuation measurements in the low temperature range. They proposed a method which achieves the desired goal in many cases, provided the Fermi surface is simply-connected. This method is as follows. The function $f(\boldsymbol{n}) = \Delta_{\min}(\boldsymbol{n})$ is determined using ultrasonic measurements. Next, a great circle $c(\boldsymbol{n})$ perpendicular to the direction of $\boldsymbol{n}$ is constructed for each point of a level curve γ of this function. The envelope of this family of circles is a level curve of the function $\Delta(\boldsymbol{n})$. Unfortunately, if the Fermi surface is multiply-connected, this procedure does not yield a unique Fermi surface, since it is impossible to determine to which of the disconnected parts the level curves belong.

2.3. Experimental Data

Anisotropy of the energy gap has been detected by the ultrasonic method in many superconducting elements. For a detailed review of the experimental data, see /129, 159/. First, investigations of superconducting tin /132–137/ should be mentioned. In Figure 21 the dependence of the absorption on the direction of the sound wave can be seen directly /133/.

Claiborne and Einspruch /138/ pointed out that the difference $\Delta_{\max} - \Delta_{\min}$ decreases with increased doping of the specimen.

According to /30/, introduction of impurities brings the superconductor increasingly closer to the isotropic model. The effects of an anisotropic energy gap, as was already pointed out above, are substantial as long as $l \gg v/\Delta_{av}$. In sufficiently pure superconductors, satisfying Anderson's condition, a dependence of the critical temperature on the concentration of nonmagnetic impurity is also observed. One can therefore associate the observed measurements of the energy gap with changes in the critical temperature.

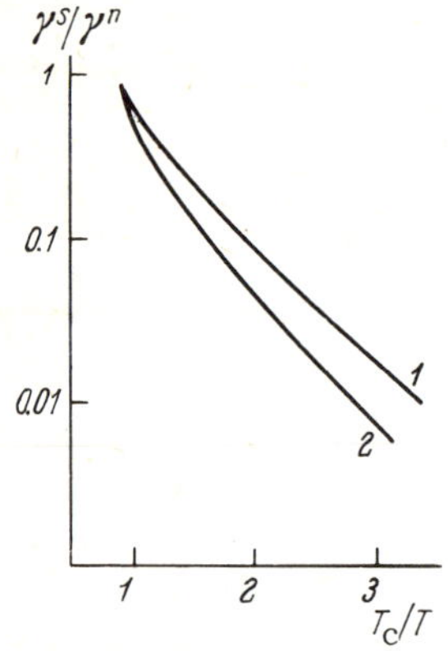

FIGURE 21. Dependence of absorption on the direction of the sound wave /133/: 1) — $q \parallel$ [001]; 2) — $q \parallel$ [110].

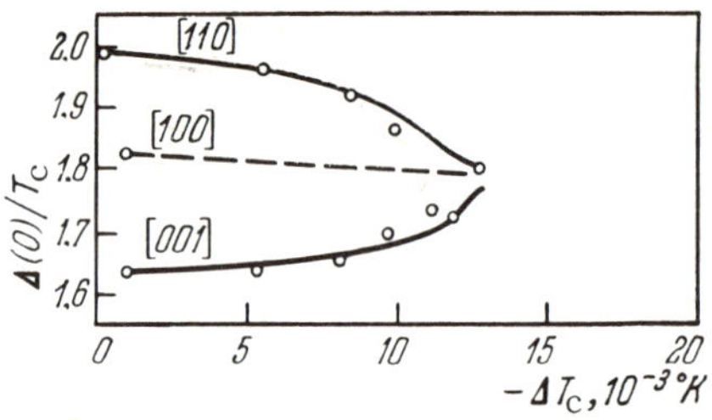

FIGURE 22. Isotropization of energy gap with increase in impurity concentration /138/.

Figure 22 illustrates the energy gap in different directions as a function of ΔT_c, as obtained in Claiborne and Einspruch /138/. It can be seen that as the impurity concentration increases the energy gap becomes increasingly isotropic and the ratio $\Delta(0)/T_c$ approaches the value predicted by the usual superconductivity theory. This is one confirmation of Anderson's theory.

According to /134, 135/, the minimum energy gap $2\Delta_{min} = 3.9\ T_c$ referred to the zone $\boldsymbol{qv} = 0$ on the Fermi surface corresponds to the orientation $\boldsymbol{q} \perp$ [101] relative to the crystallographic axes. But if, for example, $\boldsymbol{q} \perp$ [112], then $2\Delta_{min} = 4.4\ T_c$, while the $\boldsymbol{q} \perp$ [111] corresponds to $2\Delta_{min} = 4.8\ T_c$. These papers (see also /139/) confirm Pokrovskii's theorem /95/, which states that energy gaps referred to different crystallographic directions vary proportionally with the temperature. This has also been observed in an investigation of superconducting indium /140/.

A deviation from the simple exponential dependence of the attenuation coefficient has been observed at very low temperatures /141–143/. This may be associated with the participation of phonons in sound absorption (see Sec. 11), which may be a factor in this temperature range in view of the small number of electron excitations.

Ultrasonic measurements reveal an anisotropic energy gap also in Ga /144/, Va and Zn /141/, Ta /145, 146/, Re /147/, In /148—150/, Nb /151/, and in many superconductors (see, for example, /129/). In Zn, $2\Delta_{min} = 3.4\ T_c$, and $2\ \Delta_{max} = 3.8\ T_c$. In Nb, $2\Delta = 3.6\ T_c$ corresponds to $\boldsymbol{q} \parallel [100]$, and $2\ \Delta = 3.4\ T_c$ corresponds to $\boldsymbol{q} \parallel [111]$.

In /152/ the energy spectrum of Nb was investigated by measuring its specific heat in the low temperature range. According to the data in this paper, pure superconducting Nb is characterized by two energy gaps, $2\Delta_1 = 3.5\ T_c$ and $2\Delta_2 = 0.3\ T_c$. Thus, investigation of these properties by different methods leads to different values of Δ_{min} and Δ_{max}. The reason is that specific heat measurements reveal a basically multiband superconductor structure, whereas the major factor in the dependence of ultrasonic absorption on direction is the anisotropy of the energy gap.

3. ULTRASONIC ABSORPTION IN STRONG COUPLING SUPERCONDUCTORS

In a series of papers /56, 175—182/ ultrasonic absorption in strong coupling superconductors was experimentally investigated. In /175/ it is pointed out that the rate of ultrasonic attenuation in superconducting lead below T_c is significantly higher than it should be according to the usual superconductivity theory. A similar phenomenon is noted in /176, 177/. In /178, 179/ ultrasonic attenuation was investigated in polycrystalline specimens of Hg. In /180/, which also discusses sound attenuation in mercury, both pure specimens and crystals of Hg containing Cd impurities were studied. Sound absorption in superconducting Nb was considered in /181/, superconducting In in /182/. The attenuation of surface waves in Pb and In films was investigated in /156/. In all these papers the decrease of the function $\gamma^s(T)$ when temperature fell below T_c was sharper than expected from the BCS theory.

The anisotropy effect and the overlapping bands influence the temperature dependence of the attenuation coefficient. It is clear, however, that these effects are an important factor in the low temperature range, far from T_c. Another very important case should be considered. If the specimen under study is not sufficiently pure (does not satisfy the condition $l \gg \xi_0$, where l is the free path), i. e., is not prepared, the effects of anisotropy and multiband structure are not a factor. In the above-mentioned papers sound absorption was usually studied in moderately doped specimens, and therefore the above property of the temperature dependence of the attenuation coefficient cannot be explained in terms of anisotropy or overlapping bands.

If the specimen is fairly pure, however, the effects of anisotropy and multiband structure naturally exert a substantial influence on the temperature dependence of the attenuation coefficient. References /177, 179/ discuss the possible role of multiband structure in Pb and Hg, respectively. However, as mentioned in Chapter I, Sec. 9, the effects associated with the presence of overlapping bands can be isolated unambiguously by introducing impurities.

The above property of the absorption process is associated with the anomalous character of lead, mercury and bismuth, which are superconductors with strong electron-phonon interactions. It can be explained /183/ (see also /184/) on the basis of the strong coupling theory.

Absorption is described by the imaginary part of the polarization operator (3.10), where

$$G=\frac{u^2}{\varepsilon-i\omega_n}-\frac{v^2}{\varepsilon+i\omega_n},$$

$$\left.\begin{matrix}u^2\\ v^2\end{matrix}\right\}=\frac{1}{2}\left(1\pm\frac{\xi}{\varepsilon}\right),\quad \varepsilon=\sqrt{\xi^2+\Sigma^2(\omega_n)},$$

$$F=\frac{\Sigma(\omega_n, T)}{2\varepsilon}\left[\frac{1}{\varepsilon-i\omega_n}+\frac{1}{\varepsilon+i\omega_n}\right].$$

Note that here the self-energy part $\Sigma(\omega_n, T)$ figuring in the Green's function depends essentially on the "frequency" ω_n ($\omega_n=(2n+1)\times \pi T$). According to /84, 85/ (see Chapter I, Sec. 8),

$$\Sigma(\omega_n, T)=C(T)\frac{\omega^2}{\omega^2+\omega_n^2}+\Sigma' \qquad (\Sigma'\ll\Sigma),$$

where ω is the effective phonon frequency.

We next transform (3.10) to an integral in the complex plane (a similar method was used by Pokrovskii to calculate the ultrasonic absorption coefficient in anisotropic superconductors; see Sec. 2) by the substitution

$$T\Sigma\to\frac{i}{4\pi}\int_\Gamma d\omega\,\mathrm{th}\,(\omega/2T)$$

(where the contour Γ consists of two straight lines parallel to the imaginary axis). Next, deforming the integration contour, using residues and the substitution $i\omega_n\to\omega+i\delta$, we obtain

$$\mathrm{Im}\,\Pi=\mathrm{const}\int dp\left\{\left(1+\frac{\xi\xi'-\Delta^2}{\varepsilon\varepsilon'}\right)(f-f')\,\delta(\varepsilon'-\varepsilon-\omega)+\right.$$
$$\left.+\left(1-\frac{\xi\xi'-\Delta^2}{\varepsilon\varepsilon'}\right)(1-f-f')\,\delta(\varepsilon'+\varepsilon-\omega)\right\}. \tag{3.14}$$

The quantity Δ in (3.14) is the solution of the equation

$$\omega = \Sigma\,(-i\omega)$$

and represents the energy gap in the spectrum, as determined for $T \to T_c$ by (1.21). A similar expression for the absorption coefficient is obtained in the usual case /20, 32/ (see Sec. 1).

Integrating over angles and energy, we obtain the formulas (3.2) and (3.3) of the usual theory. However, the energy gap figuring therein is now that corresponding to anomalous superconductors. Whereas in the usual case the coefficient α in the equation

$$(\Delta/T)_{T\to T_c} = \alpha\,(1 - T/T_c)^{1/2}$$

equals $\alpha = 3.06$; for Pb, however, $\alpha = 4$ /84, 85/. The latter value also indicates a more abrupt decrease in ultrasonic attenuation as the temperature falls below T_c.

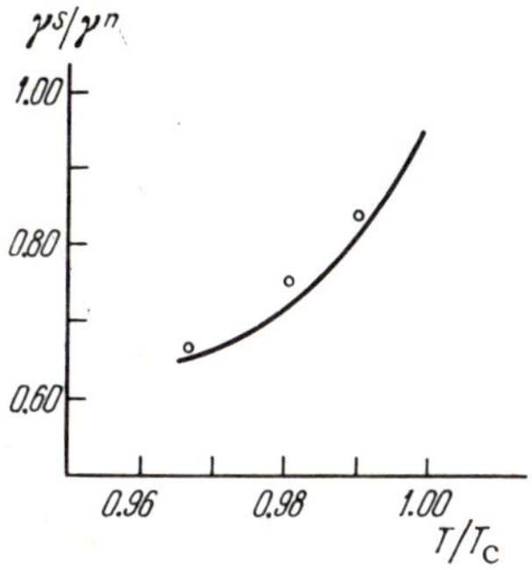

FIGURE 23. Ultrasonic attenuation coefficient in strong coupling superconductors. The experimental data refer to Pb /175/.

Figure 23 presents experimental data /175/ corresponding to measurements near T_c and a theoretical curve /183/ based on the above considerations. It can be seen that equation (3.3) with the function $\Delta/T_{T\to T_c}$ for lead describes the experimental data of /175/ quite well. The experimental data obtained in /181/ in an investigation of ultrasonic attenuation in Nb also indicate, as the authors point out, a larger value of the coefficient α than in the BCS theory.

4. ULTRASONIC ABSORPTION IN SUPERCONDUCTORS WITH OVERLAPPING BANDS

4.1. The Absorption Coefficient

The properties of superconductors with overlapping bands were discussed in the first chapter. Multiband structure should also be taken into account in an investigation of sound attenuation processes. Furthermore, as we shall see in the sequel, experimental data obtained from ultrasonic measurements may also be used to determine parameters characterizing multiband superconductors.

We proceed with the calculation of the attenuation coefficient /28/. The Hamiltonian has the form

$$H = \sum_{i,\,k,\,q,\,k'=k+q} g^{i}_{kk'} a^{i+}_{k'} a^{i+}_{k} b_q .$$

Processes in which an electron, having absorbed a sound quantum, undergoes a transition from one band to another are not considered, since such transitions are accompanied by a change in electron momentum of the order of p_0, and this is impossible at the frequencies considered, which satisfy the condition $\omega \lesssim T, \Delta^i(0)$ Here one expects the ratio γ^s/γ^n to be expressible as a sum of terms each of which contains the factor $[\exp(\Delta^i/T) + 1]^{-1}$, as in the usual theory (see Sec. 1) /121/.

To calculate the attenuation coefficient, which is determined by the imaginary part of the phonon Green's function, we write the expression for the sum $\sum_i \Pi^i$ of polarization operators, where

$$\Pi^i = \frac{T}{(2\pi)^3} \sum_{\omega_{n'}} \int dk' g^{i2}_{kk'} \left[G^i(k')G^i(k'-q) - F^i(k')F^i(k'-q)\right], \quad (3.15)$$

$$G^i = \frac{u^{i2}_k}{\varepsilon^i_p - \omega_n} - \frac{v^{i2}_k}{\varepsilon^i_p + \omega_n}; \quad u^2, v^2 = \frac{1}{2}\left(1 \pm \frac{\xi_p}{\varepsilon_p}\right),$$

$$F^i = \frac{\Delta^i}{2\varepsilon^i}\left[(\varepsilon^i - i\omega_n)^{-1} + (\varepsilon^i + i\omega_n)^{-1}\right].$$

Changing over now to integration in the complex plane (see Sec. 3) in (3.15), we obtain the following expression for $\gamma^i \sim \operatorname{Im}\Pi^i$:

$$\gamma^i \sim \int g^{i2}_{kk'} dp \left\{\left(1 + \frac{\xi\xi' - \Delta^{i2}}{\varepsilon^i \varepsilon^{i'}}\right)(f - f')\,\delta(\varepsilon' - \varepsilon - \omega) + \right.$$
$$\left. + \left(1 - \frac{\xi\xi' - \Delta^{i2}}{\varepsilon^i \varepsilon^{i'}}\right)(1 - f - f')\,\delta(\varepsilon' + \varepsilon - \omega)\right\}^i . \quad (3.16)$$

An alternative derivation of this formula is to go over to the new Fermi operators, using (1.30), in the original Hamiltonian, and then to write down the probability of absorption of a sound quantum and the probability of the reverse process. The absorption coefficient in the usual single band case is calculated in this way (see Sec. 1). The result is the same formula (3.16). We next integrate over the angle and obtain the following result for the absorption coefficient:

$$\gamma^s \sim \sum_i |g^i_{kk'}|^2 \, [(m^i)^2/q] \, F(\omega/T, b^i),$$

where the function $F(x)$ is given by the right-hand side of equation (3.2).

In the most interesting case, $x = \omega/T \ll 1$, we obtain the following equation for the ratio of the ultrasonic absorption coefficients in the superconducting and normal states:

$$\frac{\gamma^s}{\gamma^n} = \sum_i \frac{\lambda^i}{e^{\Delta^i/T} + 1}, \qquad (3.17),$$

where

$$\lambda^i = \frac{2\,|g^i_{kk'}|^2_F \,(m^i)^2}{\sum_i |g^i_{kk'}|^2_F \,(m^i)^2}. \qquad (3.17').$$

Thus, we see that the temperature dependence of each of the terms in equation (3.17) has the same form as in the usual case. The relative contribution of each of the bands depends not only on Δ^i, but also by the values of the coefficients λ^i. It is essential that these coefficients be expressed in terms of the corresponding matrix elements.

If the model is such that the matrix elements g^i are identical for different bands, the result is $\lambda^i = 2\,(m^i)^2/\sum_i (m^i)^2$, as derived by the method of correlation functions in /321/.

When $l \lesssim \xi_0$, the temperature dependence of γ^s/γ^n is described by equation (3.3), although the value of the energy gap is determined by the normal parameters of the overlapping bands /323/.

It is clear from (3.17) that when $T \to 0$, the main contribution to the absorption is made by the band with the smallest energy gap. This gives rise to a deviation from the simple exponential dependence and a certain retardation in the decrease of the function $\gamma^s(T)$. Several authors have noticed this deviation of the temperature dependence of the attenuation coefficient from its usual exponential form as $T \to 0$. Thus, in /141/ this phenomenon was observed in vanadium. A similar retardation has been noted in superconducting tin /138/. According to /30/, introduction of impurities gradually eliminates the multiband effects, as can be seen in Figure 24, which is taken from /138/.

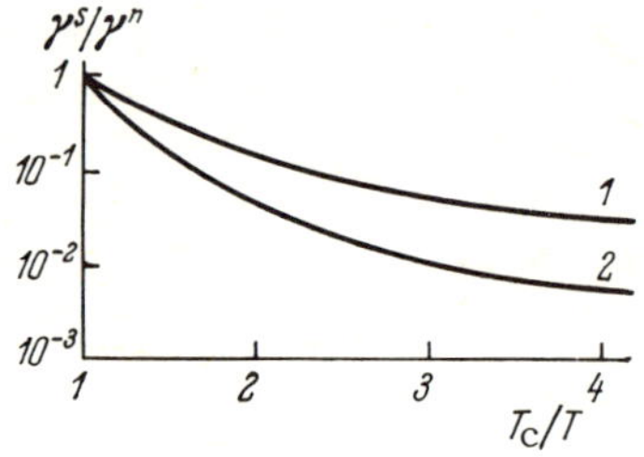

FIGURE 24. Sound attenuation in multiband superconductors.

Curve 1: pure Sn; curve 2: Sn + 0.1 at.% In. In pure superconductors the decrease in γ^s/γ^n is slower.

In /173/ a decrease in γ^s/γ^n, more abrupt than in the usual case, was observed in pure Nb as $T \to T_c$. With further decrease in temperature, however, this anomaly begins to disappear, the decrease becomes slower, and the points corresponding to the experimental data approach the curve of the usual BCS theory. This behavior of γ^s/γ^n in Nb may be due to the strong electron-phonon coupling in this superconductor. This coupling also results (see Sec. 3) in a sharper drop in the attenuation coefficient near T_c. The subsequent retardation in the temperature dependence may be associated with the multiband structure of pure Nb.

Note that anisotropy causes slower changes in the energy gap than does the effect of overlapping bands. It is noteworthy that the same reference /138/ cites data for the energy gaps Δ^1 and Δ^2 ($2\Delta^1/T_c = 2.8$ and $2\Delta^2/T_c = 8$), which apparently belong to different energy bands (see the data for Nb in Chapter I, Sec. 9). In /129/ it is pointed out that one of the directions of the sound wave relative to the axes exhibits a deviation from exponential dependence. We note in this context that the appearance of a multiband structure in Sn in tunneling experiments has been pointed out in /164/.

It is evident from equation (3.17) that the coefficients λ^i which together with Δ^i determine the contributions of the various bands, depend essentially on the matrix elements $g^i_{kk'}$, which describe the interaction of sound phonons with electrons of the various bands. To determine how the coefficients λ^i depend on the parameters of the corresponding bands involves calculation of the matrix elements $g^i_{kk'}$; the problem is therefore extremely complicated. Above (Chapter I, paragraph 9.2) we presented expressions calculated by the Wigner-Seitz method in the rigid ion approximation, which may be used for the estimates.

4.2. Determination of the Basic Parameters of Multiband Superconductors

Ultrasonic measurements are a highly effective tool for determination of the parameters describing the properties of superconductors with overlapping bands.

In Chapter I, Sec. 9, we stated that multiband superconductors are characterized by a larger number of constants $\rho_{i\nu}$ than in the usual case, and therefore the critical temperature cannot be the only parameter determining their properties. Various electromagnetic, thermal, kinetic, and other properties of superconductors with overlapping bands are described by equations in which the values of the energy gaps $\Delta^i(0)$ appear along with T_c. When $T \to T_c$ the coefficients α^i, which determine the values of the energy gaps $\Delta^i(T)_{T\to T_c} = \alpha^i[1 - T/T_c]^{1/2}$, appear in the equations. Since $\eta_0^i = \Delta^i(0)/\Delta^1(0) = \eta_c^i = (\Delta^i/\Delta^1)_{T\to T_c}$, it is sufficient to determine the values of T_c, $\Delta^i(0)$, and each energy gap for $T \to T_c$ in order to describe the properties of superconductors with a multiband structure. Thus, when using the two-band model, one must know $\Delta^1(0), \Delta^2(0)$, and α^1 (Δ^1 and Δ^2 are the energy gaps in the first and second bands). To determine these quantities, ultrasonic measurements can be used together with the thermal measurements.

When $T \to 0$ the minimum energy gap makes the main contribution to attenuation (as it does to the lattice thermal conductivity — see Chapter I, Section 9.3 — and to specific heat /152, 171/). In this temperature range $\gamma^s/\gamma^n \sim \lambda^1 \exp(-\Delta^1(0)/T)$ (we assume $\Delta^1 < \Delta^2$). It is clear that experimental measurements of γ^s/γ^n yield values for λ^1 and $\Delta^1(0)$ (it is naturally most convenient to use a logarithmic scale; $\Delta^1(0)$ is then the slope of a suitable straight line). The value of $\Delta^1(0)$ may also be determined directly by measuring $\varkappa_{pe}^s$ (see Chapter I, Section 9.3) or the electronic specific heat /152, 160/. The ratio of the electron-phonon interaction constants for a metal with overlapping bands can also be determined from these measurements. The value of $\Delta^2(0)$ can then be determined without difficulty. The value of λ^2 is determined from the equality $\sum_{i=1,2} \lambda^i = 2$.

The ratio $(\gamma^s/\gamma^n)_{T\to T_c}$ can be expressed as

$$\left(\frac{\gamma^s}{\gamma^n}\right)_{T\to T_c} = \sum_i \frac{\lambda^i}{e^{\eta^i \Delta^1/T} + 1},$$

where $\eta^i = \Delta^i/\Delta^1$. The values of λ^i may be assumed known, since they are measured for $T \to 0$. The same holds true for the values of η^i, since, as already mentioned (see Chapter I, Sec. 9.1), $\eta^i_{T=0} = \eta^i_{T=T_c}$. Thus, by measuring $(\gamma^s/\gamma^n)_{T\to T_c}$, one can also evaluate the parameter $\Delta^1_{T\to T_c}$ and thereby all the energy gaps at $T \to T_c$. Determination of the parameters $\Delta^i(0)$, T_c and $\Delta^i_{T\to T_c}$ makes it possible to describe the basic properties of superconductors with overlapping bands (see Chapter I, Sec. 9).

It follows from the foregoing discussion that experiments to determine the parameters of superconductors with overlapping bands by ultrasonic measurements should be of considerable interest.

5. THRESHOLD PHENOMENA

The presence of an energy gap in the energy spectrum makes it possible to observe the so-called threshold effects in superconductors /118, 185, 186/ (see Chapter I, Sec. 4 and Chapter II, Sec. 1).

While a phonon of any frequency can be absorbed in a normal metal, in a superconductor at $T = 0$ ultrasound will be attenuated only when $\omega \geqslant 2\Delta(0)$. The corresponding process of the creation of an electron excitation pair by a phonon at finite temperatures is possible only when the phonon energy exceeds the threshold value $\omega = 2\Delta(T)$. We have already considered the contribution of this process in our discussion of heat conduction (see (1.12)) and the sound absorption coefficient (see (3.1) (3.2)). The terms containing the step function $D(x)$ describe the contribution of the corresponding threshold effects to the temperature dependence of the kinetic coefficients.

We now consider the threshold effect at $T = 0$, which was studied by Pokrovskii in /185/. In order to determine the coefficient of ultrasonic attenuation due to decay of a phonon into an electron excitation pair $(\omega_{\boldsymbol{q}} = \varepsilon_{\boldsymbol{p}} + \varepsilon_{\boldsymbol{p}-\boldsymbol{q}})$, we must calculate the polarization operator

$$\Pi(\boldsymbol{q}, \omega) = \Pi_1 + \Pi_2,$$

where (see (3.10))

$$\Pi_1(\boldsymbol{q}, \omega) = \frac{i}{(2\pi)^4}\int d^4p \left[\frac{u^2_{\boldsymbol{p}}}{\varepsilon_{\boldsymbol{p}} - \eta - i\delta} - \frac{v^2_{\boldsymbol{p}}}{\varepsilon_{\boldsymbol{p}} + \eta - i\delta}\right] \times \\ \times \left[\frac{u^2_{\boldsymbol{p}-\boldsymbol{q}}}{\varepsilon_{\boldsymbol{p}-\boldsymbol{q}} - \eta + \omega - i\delta} - \frac{v^2_{\boldsymbol{p}-\boldsymbol{q}}}{\varepsilon_{\boldsymbol{p}-\boldsymbol{q}} + \eta - \omega - i\delta}\right],$$

$$\Pi_2(\boldsymbol{q}, \omega) = \frac{\Delta^2}{i(2\pi)^4}\int d^4p\, [(\varepsilon_{\boldsymbol{p}} - \eta - i\delta)(\varepsilon_{\boldsymbol{p}} + \eta - i\delta) \times \\ \times (\varepsilon_{\boldsymbol{p}-\boldsymbol{q}} - \eta + \omega - i\delta)(\varepsilon_{\boldsymbol{p}-\boldsymbol{q}} + \eta - \omega - i\delta)]^{-1}.$$

Integrating over η and then transforming to integration over ε and ε_1 $(\xi_1 = \xi_{p-q})$, we obtain

$$\operatorname{Im}\Pi(q, \omega) = \frac{1}{4\pi q}\int_{\Delta}^{\infty} d\varepsilon \int_{\Delta}^{\infty} d\varepsilon_1 \frac{\varepsilon\varepsilon_1 + \Delta^2}{[(\varepsilon^2 - \Delta^2)(\varepsilon_1^2 - \Delta^2)]^{1/2}}\, \delta(\varepsilon + \varepsilon_1 - |\omega|). \quad (3.18)$$

The attenuation is now determined from the equation

$$D_0^{-1} - \Pi = 0 \quad (3.19)$$

(D_0 is the phonon Green's function). It is immediately clear that, as expected, $\operatorname{Im}\Pi = 0$ when $\omega < 2\Delta$.

We next consider the threshold absorption when the frequency ω is near the threshold, i. e., $\omega = 2\Delta(1 + \alpha/2)$, $\alpha \ll 1$. In this case, by (3.18),

$$\operatorname{Im}\Pi = \Delta/4q,$$

which together with (3.19) yields the attenuation

$$\operatorname{Im}\omega_q = \frac{\lambda_0\pi^2}{4}\cdot\frac{\Delta}{M^{1/2}}. \quad (3.20)$$

Thus, as follows from (3.20), the absorption at the threshold point jumps from zero to a finite value. This jump, which is determined by the ratio $\operatorname{Im}\omega_q/\omega_q$, is of the order of 10^{-2}. The spread of this jump in attenuation is due to the logarithmic singularity of $\operatorname{Re}\Pi$ and occurs in a very small frequency range.

In $\omega \gg 2\Delta$, one naturally obtains the equation $\operatorname{Im}\Pi = |\omega|/4\pi q$ /187/, describing process of creation of an electron hole pair by a phonon in a normal metal.

As expected, in the anisotropic case /186/ the threshold frequency ω_0 is determined by the minimum of the energy gap near the curve $\boldsymbol{qv} = 0$. Thus, we see that the threshold frequency ω_0 depends on the direction of the sound propagation. It is determined by the minimum of the function $\Omega(\vartheta, \varphi)$, which is

$$\Omega^2(\vartheta, \varphi) = 4\Delta^2(\vartheta, \varphi)/(1 - v^2c^{-2}\cos^2\vartheta)$$

(ϑ is the angle between the vectors $\boldsymbol{q}$ and $\boldsymbol{v}$). Calculation of the polarization operator allowing for anisotropy leads to the following absorption-frequency relationship near the threshold ($\omega^2 = \omega_0^2(1 + \alpha)$; $\alpha \ll 1$):

$$\operatorname{Im}\Pi \sim (\omega - \omega_0)^{1/2}.$$

One sees that here, near the absorption threshold $\gamma^s(\boldsymbol{n})$ is described by the law $\gamma^s(\boldsymbol{n}) \sim (\omega_0/v)(\omega - \omega_0)^{1/2}$.

As mentioned above (see Sec. 1), at finite temperatures threshold effects are significant at frequencies $\omega \geqslant 2\Delta(T)$ /118/. Thus, even low-energy sound quanta may decay into an electron excitation pair at near-critical temperatures.

In the paper /118/ just cited, Privorotskii calculates the derivative $d\gamma^s/dT$. According to (3.2) and (3.3), this derivative goes to infinity at the critical point. However, calculation of the threshold absorption based directly on formula (3.1) yields a finite value. The point is that, when considering such fine effects as the jumps in the derivative $d\gamma^s/dT$ (see below) at $T = T_c$ and the absorption coefficient, the approximation (1.13) cannot really be adopted in equation (3.1), since it is not accurate enough to detect such small effects. For example, when $\omega < (u/v_F)\, T_c$, the value of the derivative $(d\gamma^s/dT)_{T=T_c}$ turns out to be

$$\left(\frac{d\gamma^s}{dT}\right)_{T=T_c} = \frac{1}{3}\gamma^n \frac{c^2}{\omega T_c}\left(\frac{u}{v_F}\right)^2, \quad c = 3.05 T_c^{1/2}.$$

In a normal metal the ultrasonic absorption coefficient is independent of temperature /154/, i. e., the derivative vanishes. Consequently, the transition from normal to superconducting state involves a jump in the derivative $d\gamma^s/dT$. However, even at frequencies such as 10^8 sec^{-1} it is very small: $(T_c/\gamma^n)\,(d\gamma^s/dT) \sim 10^{-4}$. Remarkably enough, at high frequencies the derivative $d\gamma^s/dT$ is negative, implying that the absorption coefficient increases as the temperature falls below T_c.

The sound absorption coefficient of a sufficiently high frequency may increase when the temperature falls to a value T_{thr}, determined by the equality $\omega = 2\Delta(T_{thr})$. This rise in the absorption of hypersound ($\omega \sim 10^{10}$ Hz) in superconducting indium is reported in /188/. The jump in the absorption coefficient at $\omega = 2\Delta$ is

$$\delta\gamma/\gamma^n = {}^1/_2\pi\,[1 - 2f(\omega/2)]$$

(the Fermi function $f(\omega/2)$ is calculated at a temperature T_{thr}, such that $\omega = 2\Delta(T_{thr})$). In /189/ a jump in sound absorption in Al and Ir was detected.

Another peculiar threshold effect, discussed by Pokrovskii /185/, is also possible in superconductors: radiation of a phonon by an electron excitation. In a normal metal this phenomenon (the corresponding conservation law has the form $\varepsilon_p = \varepsilon_{p-q} + \omega_q$) is always possible, since the Fermi velocity is greater than the velocity of sound. In superconductors, however, the velocity of electron excitations on the Fermi surface is zero. Therefore, radiation of a phonon is possible in superconductors only beginning from a certain value of the electron momentum $p_1 = p_F + (\Delta/p_F)(u/v_F)$, at which the velocity of the electron excitations becomes equal to the velocity of sound. It is this circumstance that gives the phenomenon its threshold character.

The phonon radiation also results in an attenuation of electron excitations, whose magnitude may be determined by calculating the self-energy part $\Sigma_1(\boldsymbol{p}, \eta)$:

$$\Sigma_1(\boldsymbol{p}, \eta) = \frac{1}{i(2\pi)^4} \int G_0(\boldsymbol{p} - \boldsymbol{q}) D(\boldsymbol{q}) \, d^4q.$$

The imaginary part Im Σ_1, which describes the attenuation, is

$$\operatorname{Im} \Sigma_1 = \frac{\lambda_0 \pi}{4pp_F} \int_0^{q_m} q\omega_q \, dq \int_{\xi_{p-q}}^{\xi_{p+q}} \delta(\varepsilon_{\xi_1} + \omega_q - |\eta|) \, d\xi_1. \qquad (3.21).$$

Evaluation of the integral in (3.21) yields the following value of the attenuation near the threshold $\eta = \eta_0(1 + \alpha)$, $\alpha \ll u^2/v_F^2$ (η_0 is the threshold value, $\eta_0 \approx \Delta(1 + u^2/2v_F^2)$):

$$\operatorname{Im} \Sigma_1(p_F, \eta) = \frac{\lambda_0 \pi}{12\sqrt{2}} \frac{\Delta^2}{\omega_0^2} \Delta \alpha^3. \qquad (3.21')$$

6. CHANGE IN THE VELOCITY OF SOUND

The effect of the superconducting transition on the phonon spectrum has been investigated in a series of theoretical /190–193/ and experimental /140, 194–198/ papers. In this connection we must consider the phonon Green's function in a superconductor. The polarization operator appearing in this Green's function contains, apart from quantities characterizing a normal metal, other terms containing an F-function. These are naturally proportional to Δ^2. Therefore, during transition through T_c from normal to superconducting state, there are no jumpwise changes in the phonon spectrum or, correspondingly, in the velocity of sound.

The corrections increase with increasing distance from the critical temperature and are maximal at $T = 0$. In the intermediate temperature range, they have values of the same order as at $T = 0$, since there the energy gap is only weakly dependent on temperature. However, even at $T = 0$ the change in the phonon spectrum caused by the superconducting state is also very small, being described by a quantity of the order of $(\Delta(0)/\varepsilon_F)^2$ /191/. Small effects of this kind can be observed experimentally (excluding, of course, temperatures very close to T_c). The fact that the velocity change is small has been confirmed in experimental papers /140, 194—196/. Usually, $\Delta v/v \sim 10^{-6}$. A change of the order of 10^{-4} has been observed only in Pb /197, 198/, but of course this is also very small.

In principle, marked anomalies in the sound dispersion law can be observed in superconducting semiconductors or semimetals, where the quantity ε_F can be smaller than it is in metals. At frequencies $\omega \sim T_c$ and sound propagation velocity $u \sim v_F$ the wavelength of sound becomes comparable in order of magnitude to the size of a Cooper pair (in fact, under these conditions $\lambda/\xi_0 \approx \approx uT_c/v_F\omega \sim 1$). In this case (theoretically investigated in /193/), marked changes can be expected in the phonon spectrum of a superconductor as compared with a normal metal.

7. SCATTERING WITH CHANGE OF FREQUENCY

In both normal and superconducting metals, apart from absorption of a sound wave, there may be a scattering process with a frequency change (increase or decrease). This type of sound scattering is clearly determined by electron-phonon interaction. If the interaction energy of electrons with coordinates $\boldsymbol{r}_i$ and ions with coordinates $\boldsymbol{R}_i$ is $H_{ei}(\boldsymbol{r}, \boldsymbol{R})$ the electron-phonon interaction (as can be shown by adiabatic perturbation theory) is determined by the difference $H_{ei}(\boldsymbol{r}, \boldsymbol{R}) - H_{ei}(\boldsymbol{r}, \boldsymbol{R}_0)$, where $\boldsymbol{R}_0$ are the equilibrium coordinates of the ions (see /97/). In our case we can limit ourselves to second-order terms in the expansion in powers of $\Delta\boldsymbol{R} = \boldsymbol{R} - \boldsymbol{R}_0$:

$$H' = H_1 + H_2$$

$$H_1 = \sum_{i,\,\alpha}^{N,3} \frac{\partial H_{ei}}{\partial X_{i\alpha}}\bigg|_{\boldsymbol{R}=\boldsymbol{R}_0} \Delta X_{i\alpha},$$

$$H_2 = \sum_{\substack{i,\,\alpha\\ j,\,\beta}} \frac{\partial^2 H_{ei}}{\partial X_{i\alpha}\partial X_{j\beta}}\bigg|_{\boldsymbol{R}=\boldsymbol{R}_0} \Delta X_{i\alpha}\Delta X_{j\beta},$$

$\Delta X_{i\alpha}$ is expressed as usual in terms of the phonon operators $b_{q\lambda}$:

$$\Delta X_{i\alpha} = \sum_{q,\,\lambda} \frac{1}{\sqrt{2NM\omega_{q\lambda}}} e_{q\lambda\alpha}\,[b_{q\lambda}\exp(i\boldsymbol{q},\,\boldsymbol{R}_{0i}) + \text{h. c.}];\ \hbar = 1,$$

where $\boldsymbol{q}$ is the phonon vector, $\omega_{q\lambda}$ its frequency, $\boldsymbol{e}_{q\lambda}$ the phonon polarization unit vector, λ the polarization index, and M the mass of an ion.

The sound scattering process under discussion is determined by the term H_1 in the second approximation of perturbation theory and by H_2 in the first approximation. The sound absorption process considered previously is clearly determined by the term H_1 in the first approximation of perturbation theory. It is readily seen that the collision cross section of this type of scattering significantly below the Debye temperature is smaller than the sound absorption cross section by a factor of $\sqrt{m/M} \sim 10^{-2}$ (m is the electron mass). Thus, the probability of the process is quite high, and it can easily be observed experimentally.

Apart from the scattering by electrons, sound may also be scattered by the lattice, owing to the third- and fourth-order anharmonic terms in the lattice vibration energy. However, comparison of the matrix elements of H_1 and H_2 with the matrix elements of the anharmonic terms shows that the probability of sound waves being scattered by the lattice is usually less than for scattering by electrons (in the low temperature range under discussion). Scattering of sound in metals by the lattice may therefore be neglected in the high frequency range.

The probability of the scattering of a sound wave having frequency ω_1, momentum $\boldsymbol{k}_1$, and polarization λ_1, with the emission of phonons having frequency ω_2, momentum $\boldsymbol{k}_2$, and polarization λ_2 is determined for a normal metal by the expression

$$\begin{aligned} d\boldsymbol{k}_2 \int & N_{\boldsymbol{k}_1\lambda_1}(N_{\boldsymbol{k}_2\lambda_2}+1) f(\boldsymbol{p}_1)[1-f(\boldsymbol{p}_2)] \times \\ & \times \Bigg| \Bigg[M_{\boldsymbol{p}_1,\boldsymbol{k}_1,\lambda_1} M_{\boldsymbol{p}+\boldsymbol{k}_1,-\boldsymbol{k}_2,\lambda_2} \frac{1}{\omega_1+\xi_{\boldsymbol{p}_1}-\xi_{\boldsymbol{p}_1+\boldsymbol{k}_1}+i\gamma_1} + \\ & + M_{\boldsymbol{p}_1,-\boldsymbol{k}_2,\lambda_1} M_{\boldsymbol{p}_1-\boldsymbol{k}_2,\boldsymbol{k}_1,\lambda_2} \frac{1}{-\omega_2+\xi_{\boldsymbol{p}_1}-\xi_{\boldsymbol{p}_1-\boldsymbol{k}_2}+i\gamma_2} + \\ & + L_{\boldsymbol{p}_1,\boldsymbol{p}_2,\boldsymbol{k}_1,\lambda_1,\boldsymbol{k}_2,\lambda_2} \Bigg] \Bigg|^2 \delta(\xi_{\boldsymbol{p}_1}+\omega_1-\xi_{\boldsymbol{p}_2}-\omega_2)\, d\boldsymbol{p}_1, \end{aligned} \tag{3.22}$$

and for a superconductor by the following expression, derived using the Hamiltonian H' of (1.7) (for simplicity we omit the terms associated with the creation of virtual electron pairs):

$$\begin{aligned} d\boldsymbol{k}_2 \int & N_{\boldsymbol{k}_1\lambda_1}(N_{\boldsymbol{k}_2\lambda_2}+1) f(\boldsymbol{p}_1)[1-f(\boldsymbol{p}_2)] \times \\ & \times \Bigg| \Bigg[M_{\boldsymbol{p}_1,\boldsymbol{k}_1,\lambda_1} M_{\boldsymbol{p}_1+\boldsymbol{k}_1,-\boldsymbol{k}_2,\lambda_2} \left(\frac{R^+_{\boldsymbol{p}_1,\boldsymbol{p}_1+\boldsymbol{k}_1} R^+_{\boldsymbol{p}_1+\boldsymbol{k}_1,\boldsymbol{p}_2}}{\varepsilon_{\boldsymbol{p}_1}-\varepsilon_{\boldsymbol{p}_1+\boldsymbol{k}_1}+\omega_1+i\gamma'_1} + \right. \\ & \left. + \frac{R^-_{\boldsymbol{p}_1,\boldsymbol{p}_1+\boldsymbol{k}_1} R^-_{\boldsymbol{p}_1+\boldsymbol{k}_1,\boldsymbol{p}_2}}{\varepsilon_{\boldsymbol{p}_1}+\varepsilon_{\boldsymbol{p}_1+\boldsymbol{k}_1}+\omega_1+i\gamma_1''} \right) + M_{\boldsymbol{p}_1,-\boldsymbol{k}_2,\lambda_1} M_{\boldsymbol{p}_1-\boldsymbol{k}_2,\boldsymbol{k}_1,\lambda_2} \times \\ & \times \left(\frac{R^+_{\boldsymbol{p}_1,\boldsymbol{p}_1-\boldsymbol{k}_2} R^+_{\boldsymbol{p}_1-\boldsymbol{k}_2,\boldsymbol{p}_2}}{\varepsilon_{\boldsymbol{p}_1}-\varepsilon_{\boldsymbol{p}_1-\boldsymbol{k}_2}-\omega_2+i\gamma_2'} + \frac{R^-_{\boldsymbol{p}_1,\boldsymbol{p}_1-\boldsymbol{k}_1} R^-_{\boldsymbol{p}_1-\boldsymbol{k}_2,\boldsymbol{p}_2}}{\varepsilon_{\boldsymbol{p}_1}+\varepsilon_{\boldsymbol{p}_1-\boldsymbol{k}_2}-\omega_2+i\gamma_2''} \right) + \\ & + L_{\boldsymbol{p}_1,\boldsymbol{p}_2,\boldsymbol{k}_1,\lambda_1,\boldsymbol{k}_2\lambda_2} R^+_{\boldsymbol{p}_1,\boldsymbol{p}_2} \Bigg] \Bigg|^2 \delta(\varepsilon_{\boldsymbol{p}_1}+\omega_1-\varepsilon_{\boldsymbol{p}_2}-\omega_2)\, d\boldsymbol{p}_1. \end{aligned} \tag{3.22'}$$

Transfer processes, for which $\boldsymbol{g} \neq 0$ in the equality $\boldsymbol{p}_1 + \boldsymbol{k}_1 = \boldsymbol{p}_2 + \boldsymbol{k}_2 + \boldsymbol{g}$, where $\boldsymbol{g}$ is the integral vector of the reciprocal lattice multiplied by 2π, and $\boldsymbol{p}_2$ is the electron momentum after the scattering, are not taken into account here. Similarly, one determines the probability of the process associated with the creation of electron pairs in superconductors. The threshold of this process is $2\Delta + \omega_2$. Subtracting the probability of the reverse process for (3.22) and (3.22'), we obtain the total scattering probability.

In (3.22) and (3.22'): $N_{k\lambda} = [\exp(\omega_{k\lambda}/T) - 1]^{-1}$; $f(\boldsymbol{p})$ is the Fermi distribution function (for a normal metal $f(\boldsymbol{p}) = [\exp(\xi_p/T) + 1]^{-1}$, and for a superconductor $f(\boldsymbol{p}) = [\exp(\varepsilon_p/T) + 1]^{-1}$); $\varepsilon_p = (\xi_p^2 + \Delta^2)^{1/2}$; $R^{\pm}_{pp'} = \frac{1}{\sqrt{2}}\left(1 \pm \frac{\xi_p \xi_{p'} - \Delta^2}{\varepsilon_p \varepsilon_{p'}}\right)^{1/2}$; γ_i are the widths; and $M_{pk\lambda}$ is the so-called Bloch matrix element, i.e., the matrix element of the operator H_1:

$$M_{pk\lambda} = \frac{1}{(2NM\omega_{k\lambda})^{1/2}} \int \psi_p^*(\boldsymbol{r}) (\boldsymbol{e}_{k\lambda}, \nabla_r V(\boldsymbol{r})) \psi_{p+k}(\boldsymbol{r})\, d\boldsymbol{r}$$

$(V(\boldsymbol{r}) = \sum_j V(\boldsymbol{r} - \boldsymbol{R}_{0j})$ is a periodic potential which acts on the electron as viewed from the ions; $\psi_k(\boldsymbol{r})$ is a single-particle electronic wave function); $L_{p_1p_2k_1\lambda_1k_2\lambda_2}$ is the matrix element of the operator H_2:

$$L_{p_1p_2} = \frac{1}{N(4M^2\omega_{k_1\lambda_1}\omega_{k_2\lambda_2})^{1/2}} \times$$
$$\times \sum_{\alpha,\beta}^{3,3} e_{k_1\lambda_1\alpha} e_{k_2\lambda_2\beta} \int \psi_{p_1}(\boldsymbol{r}) \frac{\partial^2 V(\boldsymbol{r})}{\partial x_\alpha \partial x_\beta} \psi_{p_2}(\boldsymbol{r})\, d\boldsymbol{r}.$$

In the acoustic dispersion law the products of Bloch matrix elements in both (3.22) and (3.22') are identical and are equal to $M_1 M_2 = \zeta\, 2\pi^2 \sqrt{\omega_1\omega_2}/mp_0$ (the volume V is assumed equal to unity), where p_0 is the Fermi momentum, $\zeta \lesssim 1$ is Fröhlich's dimensionless parameter.

Integrating (3.22) over $\boldsymbol{p}_1$, we obtain the scattering cross section as a function of $\boldsymbol{k}_1$, $\boldsymbol{k}_2$, and the angle ϑ between the vectors $\boldsymbol{k}_1$ and $\boldsymbol{k}_2$.

At the low temperatures under discussion, the probability of a phonon being scattered with an increase in frequency is quite low. The expression for the cross section is rather cumbersome and we shall not present it here; this cross section is quite large, in view of the resonance nature of the process (see /318/).

8. THE KINETIC EQUATION

We now consider the attenuation of sound having frequency $\omega \lesssim \tau^{-1}$. We shall view the sound field as a factor deforming the lattice. Since the deformation process is irreversible, sound energy is absorbed. The problem is reduced to solution of the corresponding kinetic equation and subsequent calculation of the dissipation function, which determines the absorption of the sound wave.

The lattice deformation results in a change in electron energy, which becomes equal /153/ to $\varepsilon = \varepsilon_0 + \varepsilon_{ik} u_{ik}$ (u_{ik} is the deformation tensor). Thus, a localized electron dispersion law is introduced; this is possible since the sound wavelength significantly exceeds the period of the crystal lattice.

An investigation of sound attenuation must distinguish between the direct deformation absorption γ_d and the Joule loss γ_J associated with the presence of deformation electric fields. The origin of both these components is in the final analysis the same: deformation of the crystal lattice. However, they are associated with contributions of different terms figuring in the kinetic equation.

The kinetic equation for the distribution function of electron excitations in the sound field has the form

$$\frac{\partial f}{\partial t} + \boldsymbol{v}\frac{\partial f}{\partial \boldsymbol{r}} + \frac{\partial f}{\partial \boldsymbol{p}} e\boldsymbol{E} = \left(\frac{\partial f}{\partial t}\right)_{\text{col}}. \tag{3.23}$$

The right side is determined by collisions of electrons with phonons and impurities. The sound wave modifies the energy of the electron excitation. Elastic collisions with impurities cannot be an energy relaxation mechanism, leading to establishment of an equilibrium. Equilibrium can only be established by inelastic electron-phonon interactions over a time of order τ_ε (τ_ε is the energy relaxation time).

Scattering off impurities determines only the momentum relaxation time τ_p and therefore contributes to absorption only via the creation of deformation electric fields and the accompanying currents. Sound absorption in normal metals has been studied in detail in several papers /153, 154, 174, 312—316/.

The cases to be considered in the calculation of γ_J are: 1) $l \gg \lambda$ 1) $l \lesssim \lambda$, where l is the free path (we are considering scattering off impurities so that the collision integral may be written in the τ-approximation; $l = \tau_p v$). It can be seen directly from (3.23) that in the first case collisions may be neglected. Indeed, the second term on the left of the kinetic equation is of the order of $(v/\lambda)\, f \gg f/\tau$.

Sound attenuation will then be described by the same equation as in the limiting case $\omega\tau \gg 1$ considered above. This is qualitatively clear, since neglect of collisions makes the problem entirely analogous to the case discussed in Sec. 1. Attenuation arises only thanks to the direct absorption process and the emission of sound quanta by electron excitations. Therefore, when $l \gg \lambda$, the ratio γ^s/γ^n is described by equations (3.2) and (3.3) in the isotropic case.

If the condition $l \gg \lambda$ is not fulfilled, the problem becomes substantially more complicated. It has been studied in detail in Tsuneto /215/, where the corresponding current and Joule loss, which in the final analysis describe the absorption of sound wave energy, were calculated. Tsuneto showed that in this case the ratio γ^s/γ^n is described by equation (3.3), which is thus valid for an arbitrary relationship between λ and the free path l in scattering of electrons off impurities.

The relationship between the Joule and direct deformation losses is highly important /315/ (see also /109/). If $\lambda_s \ll \lambda_{em}$ the contribution of the deformation electric fields may be neglected (λ_{em} is the electromagnetic wavelength in the substances). But if $\lambda_s \gg \lambda_{em}$ the electromagnetic fields contribute to the absorption to a degree comparable to the deformation losses. The ratio γ^s/γ^n, determined by the deformation electric fields, is described by the usual equation (3.3).

An investigation of direct deformation loss must allow for nonelastic electron-phonon interaction. This question will be discussed in Sec. 10.

9. SOUND ABSORPTION IN AN ANISOTROPIC SUPERCONDUCTOR IN THE PRESENCE OF IMPURITIES

In Sec. 2 we considered ultrasonic absorption in a pure anisotropic superconductor. We now calculate the absorption coefficient allowing for scattering of electron excitations by impurities /124/. The problem has been solved by Privorotskii for both a bulk superconductor and a film of thickness $d < l$ (l is the free path of the electron excitations).

We shall treat the sound field as a factor deforming the crystal lattice /153–155/. The absorption coefficient is determined by the formula $\gamma^s = TS/W$ (S is the entropy density of the elementary particle excitation gas, related in the usual manner to the distribution function $n(\boldsymbol{p}, \boldsymbol{r}, t)$; W is the energy density of the sound wave).

Writing the collision integral in the relaxation time approximation, we find

$$\gamma^s = -\frac{2}{W}\int \nu' |\chi|^2 \frac{\partial n_0}{\partial \varepsilon} \frac{d\varepsilon\, ds}{|d\varepsilon/d\boldsymbol{p}|}, \tag{3.24}$$

where $\nu' = \nu |\xi|/\varepsilon$ is the frequency of collisions of excitations with impurities /14/ (see Chapter I, Sec. 1);

$$n = n_0 - \frac{\partial n_0}{\partial \varepsilon}\chi, \qquad n_0 = \left[\exp\left(\frac{\varepsilon}{T}\right) + 1\right]^{-1};$$

$$\varepsilon(\boldsymbol{p}, \boldsymbol{r}, t) = \varepsilon(\boldsymbol{p}) + \varepsilon_1(\boldsymbol{p}, \boldsymbol{r}, t)$$

(the appearance of the increment $\varepsilon_1(\boldsymbol{p}, \boldsymbol{r}, t)$ is due to the displacement in the lattice, which describes the sound field).

Since the function $\partial n_0/\partial\varepsilon$ has a sharp maximum, we can integrate over ε in (3.24), obtaining the following expression for the absorption coefficient:

$$\gamma^s = \frac{2}{\hbar^3 W}\int \nu' |\chi|^2 n_0(\Delta) \frac{ds}{|d\varepsilon/dp|}, \tag{3.25}$$

the integration in (3.25) being performed over the Fermi surface. The function χ, which defines the perturbed distribution function, is found by solving a kinetic equation similar to that used in /154/ to solve the problem of sound absorption in a normal metal. Two cases are distinguished: 1) bulk metal or film with mirror reflecting boundaries; 2) film with diffusely scattering boundaries.

In the first case one observes a sharp anisotropy in the absorption coefficient in the most interesting range of low temperatures ($\exp\{\Delta_{\max} - \Delta_{\min})/T\} \gg 1$). The integrand in (3.25) includes two functions χ and n_0 which possess sharp maxima, the function χ having a maximum along the curve $v_x = 0$, and n_0 at the point on the Fermi surface where the energy gap Δ is minimum. Therefore, the absorption coefficient has a maximum for directions of the vector $\boldsymbol{k}$ such that the minimum energy gap is located precisely on the curve $v_x = 0$.

The magnitude of the absorption coefficient depends on the relationship between the parameters $(T/\delta\Delta)^{1/2}$ and $(ql)^{-1}$. When $(ql)^{-1} < (T/\delta\Delta)^{1/2}$ it is

$$\gamma^s_{\max}/\gamma^n \approx e^{-\Delta_{\min}/T}(T/\delta\Delta)^{1/2}.$$

In the opposite case $(ql)^{-1} > (T/\delta\Delta)^{1/2}$, we have

$$\gamma^s_{\max}/\gamma^n \approx e^{-\Delta_{\min}/T} qlT/\delta\Delta.$$

The angle dependence is most complicated in the temperature interval

$$(ql)^{-1} < (T/\delta\Delta)^{1/2}, \qquad \exp(\delta\Delta/T) > ql\,(\delta\Delta/T)^{1/2}.$$

In this case the anisotropy of the pre-exponential factor is significant.

The absorption γ^s_{film} in a film with diffusely scattering boundaries in the temperature range $\delta\Delta \lesssim T$ (this is precisely the condition for the temperature change of the energy gap to be most marked) is described by the equation

$$\gamma^s_{\text{film}} = 2n_0(\Delta_0)\,\gamma^n_{\text{film}}$$

where $\gamma^n_{\text{film}} = \gamma^n_{\text{bulk}} d/l$, and Δ_0 is the value of the energy gap at the point $v_x = v_z = 0$ on the Fermi surface. This equation can be used to express the energy gap as a function of direction and temperature.

In the isotropic case, the equations obtained in /124/ revert to the usual relationship (3.3). This is not accidental, since elastic scattering off impurities cannot be a relaxation mechanism for the electron excitation system, which changes its energy in the sound wave field. In the isotropic case, therefore, when an impurity is present, the absorption is the same as in a free electron gas (see Sec. 1).

In this connection we mention a paper /156/ in which the attenuation of surface waves with frequency 316 MHz was experimentally investigated in films of indium of thicknesses 500 Å and 1500 Å. In both cases the thickness of the film was less than the size of a Cooper pair. Since the indium films investigated in /156/ were not distinguished by the quality of their surface or their structural features, as was the case, for example, for the Al, Ag, and Be films, the free path in them is certainly not more than the thickness of the film. Thus, the condition $l \lesssim \xi_0$, which corresponds to "isotropization" of the energy gap, is fulfilled. As already noted, under such conditions scattering off impurities and boundaries does not influence the sound absorption process. The curves given in /156/ show that the temperature dependence of the attenuation in the In films studied is indeed independent of the thickness of the film.

In the anisotropic case, as in the presence of overlapping energy bands, the pattern is rather complicated, since electrons scattered by impurities may undergo a transition to another band or to another section of the Fermi surface. Mixing of the corresponding wave functions occurs. The change in the electronic states affects the pattern of absorption of the sound wave. This results in laws which differ from the equation (3.3) of the conventional BCS theory.

10. INELASTIC SCATTERING

We now consider the viscous loss caused by the direct deformation interaction (see Sec. 8).

The kinetic equation for the distribution function of electron excitations of a superconductor placed in a sound field and interacting with phonons is /20, 21/

$$\frac{\partial f}{\partial \varepsilon}\frac{\xi}{\varepsilon}\varepsilon_{ik}\dot{u}_{ik}=\sum_{q}\Big[|V|^2(u_k u_{k'}-v_k v_{k'})^2\{[f'(1-f)(N+1)-f(1-f')N]\,\delta(\varepsilon'-\varepsilon-\hbar\omega)+[f'(1-f)N-f(1-f')\times(N+1)]\,\delta(\varepsilon-\varepsilon'-\hbar\omega)\}+|V|^2(u_k v_{k'}+u_{k'}v_k)^2[N(1-f)(1-f')-(N+1)ff']\,\delta(\varepsilon'+\varepsilon-\hbar\omega)\Big] \tag{3.26}$$

(N is the number of phonons of frequency ω). We assume that $\lambda \gg v\tau_\varepsilon$, so we may neglect the second term on the left of (3.23). The collision integral naturally has the same form as in the problem of electronic specific heat of a pure superconductor (see Chapter I, Sec. 3).

We seek the distribution function in the form

$$f=f_0+g,\qquad g=\frac{\partial f_0}{\partial \varepsilon}c(\varepsilon,\Omega), \tag{3.27}$$

where $f_0=[\exp(\varepsilon/T)+1]^{-1}$, and $c(\varepsilon, \Omega)$ is a function which depends on the electron excitation energy and the angles defining their direction of motion.

The kinetic equation becomes

$$f_0^2 e^{\frac{\varepsilon}{T}}\frac{\xi}{\varepsilon}\varepsilon_{ik}\dot{u}_{ik}=\sum_{q}\Big[N_0|V|^2\left(1+\frac{\xi\xi'-\Delta^2}{\varepsilon\varepsilon'}\right)\times\{[c(\varepsilon',\Omega')-c(\varepsilon,\Omega)]\,f_0f_0'e^{\frac{\varepsilon+\hbar\omega}{T}}\,\delta(\varepsilon'-\varepsilon-\hbar\omega)+[c(\varepsilon',\Omega')-c(\varepsilon,\Omega)]\,e^{\frac{\varepsilon}{T}}f_0f_0'\delta(\varepsilon-\varepsilon'-\hbar\omega)\}-N_0|V|^2\left(1-\frac{\xi\xi'-\Delta^2}{\varepsilon\varepsilon'}\right)[c(\varepsilon',\Omega')-c(\varepsilon,\Omega)]\,f_0f_0'e^{\frac{\hbar\omega}{T}}\delta(\varepsilon'+\varepsilon-\hbar\omega)\Big]. \tag{3.28}$$

The expression on the left of (3.28) is an odd function of ξ. The collision operator is symmetric with respect to ξ. Consequently, $c(\varepsilon)$ is an odd function of ξ and must be sought in the form $c(\varepsilon,\Omega)=\varphi(\varepsilon,\Omega)\,\xi/|\xi|$. In so doing we remember that $\xi\xi'>0$ for scattering processes and $\xi\xi'<0$ for the processes of excitation pair creation (see Chapter I, Sec. 3).

The equation for the function $\varphi(\varepsilon, \Omega)$ is

$$f_0^2 e^{\frac{\varepsilon}{T}} \frac{|\xi|}{\varepsilon} \frac{\partial \xi}{\partial t} = \int_0^\infty N_0 |V|^2 \left(1 + \frac{\xi\xi' - \Delta^2}{\varepsilon\varepsilon'}\right) \times$$
$$\times [\varphi(\varepsilon', \Omega') - \varphi(\varepsilon, \Omega)] f_0' f_0 e^{\frac{\varepsilon + \hbar\omega}{T}} \delta(\varepsilon' - \varepsilon - \hbar\omega) \times$$
$$\times q^2 dq\, d\cos\vartheta\, d\varphi + \int_0^{\varepsilon - \Delta} N_0 |V|^2 \left(1 + \frac{\xi\xi' - \Delta^2}{\varepsilon\varepsilon'}\right) \times$$
$$\times [\varphi(\varepsilon', \Omega') - \varphi(\varepsilon, \Omega)] f_0 f_0' e^{\frac{\varepsilon}{T}} \delta(\varepsilon - \varepsilon' - \hbar\omega) \times$$
$$\times q^2 dq\, d\cos\vartheta\, d\varphi + \int_{\varepsilon + \Delta}^\infty N_0 |V|^2 \frac{1}{2} \left(1 - \frac{\xi\xi' - \Delta^2}{\varepsilon\varepsilon'}\right) \times$$
$$\times [\varphi(\varepsilon', \Omega') - \varphi(\varepsilon, \Omega)] f_0 f_0' e^{\frac{\hbar\omega}{T}} \delta(\varepsilon' + \varepsilon - \hbar\omega) \times$$
$$\times q^2 dq\, d\cos\vartheta\, d\varphi. \tag{3.29}$$

Since the wave vector $\boldsymbol{q}$ of a phonon is much smaller in magnitude than $|\boldsymbol{p}|$, we can expand $\varphi(\varepsilon', \Omega')$ in powers of $\boldsymbol{q}' = \boldsymbol{p} - \boldsymbol{p}^*$ ($\boldsymbol{p}^*$ is a vector directed along $\boldsymbol{p}$, of length equal to $|\boldsymbol{p}'|$). This method was used by Landau and Pomeranchuk in the conduction theory of normal metals /22/ and we have already employed it to calculate the electronic thermal conductivity of pure superconductors (see Chapter I, Sec. 3). Use of the method to calculate the attenuation coefficient of long wavelength sound is quite natural, since the kinetic equation involves the same collision integral. We expand the function $\varphi(\varepsilon', \Omega')$ as indicated (see equation (1.8'')) and then integrate over the angle ϑ (the polar axis is chosen in the direction of the vector $\boldsymbol{p}$). We approximate the factors

$$\left(1 \pm \frac{\xi\xi' - \Delta^2}{\varepsilon\varepsilon'}\right) \frac{\varepsilon\varepsilon'}{|\xi||\xi'|}$$

which arise during the integration process by 2, as in our calculation of thermal conductivity. Next, introducing the variables

$$z = \varepsilon/T, \quad x = \hbar\omega/T, \quad b = \Delta/T$$

and integrating both sides of the equation over z and the angle in the electron momentum space, we obtain

$$\int \frac{e^z}{(e^z + 1)^2} \varepsilon_{ik} \dot{u}_{ik} dz\, d\Omega =$$
$$= |V'|^2 T^3 \int_b^\infty dz \int_0^\infty [\varphi(z + x, \Omega) - \varphi(z, \Omega)] \times$$
$$\times \frac{e^{z+x} x^2 dx\, d\Omega}{(e^z + 1)(e^{z+x} + 1)(e^x - 1)} +$$

$$+|V'|^2 T^3 \int_b^\infty dz \int_0^{z-b} [\varphi(x-z,\ \Omega) - \varphi(z,\ \Omega)] \times \frac{e^z x^2 dx\, d\Omega}{(e^z+1)(e^{z-x}+1)(e^x-1)} +$$

$$+|V'|^2 T^3 \int_b^\infty dz \int_{z+b}^\infty [\varphi(x-z,\ \Omega) - \varphi(z,\Omega)] \times$$

$$\times \frac{e^z x^2 dx\, d\Omega}{(e^z+1)(e^{x-z}+1)(e^x-1)} +$$

$$+|V'|^2 \frac{T^5}{2} \int_b^\infty dz \int_0^\infty \frac{e^{z+x} x^4 dx\, d\Omega}{(e^z+1)(e^{z+x}+1)(e^x-1)} \frac{\partial^2\varphi}{\partial\Omega^2} +$$

$$+|V'|^2 \frac{T^5}{2} \int_b^\infty dz \int_0^{z-b} \frac{e^z x^4 dx\, d\Omega}{(e^z+1)(e^{z-x}+1)(e^x-1)} \frac{\partial^2\varphi}{\partial\Omega^2} +$$

$$+|V'|^2 \frac{T^5}{2} \int_b^\infty dz \int_{z+b}^\infty \frac{e^x x^4 dx\, d\Omega}{(e^z+1)(e^{x-z}+1)(e^x-1)} \frac{\partial^2\varphi}{d\Omega^2}, \tag{3.30}$$

$$|V'|^2 = |V|^2/q.$$

We next seek a solution of this equation in the form $\varphi(\varepsilon,\Omega) = \varphi_1(\Omega) + \varphi_2(\varepsilon,\ \Omega)$, where $\varphi_1 \gg \varphi_2$. The integrals containing the function φ_2 vanish, and from the resulting equality:

$$\int_b^\infty \frac{e^z}{(e^z+1)^2} \varepsilon_{ik}\dot{u}_{ik} dz\, d\Omega =$$

$$= |V'|^2 T^5 \int_0^\infty \frac{2x^4 dx}{e^x-1} \int_b^\infty \frac{dz\, d\Omega\partial^2\varphi_1/\partial\Omega^2}{(e^z+1)(e^{-z-x}+1)} +$$

$$+|V'|^2 \frac{T^5}{2} \int_{2b}^\infty \frac{x^4 dx}{e^x-1} \int_0^{x-b} \frac{e^z dx\, d\Omega\partial^2\varphi_1/\partial\Omega^2}{(e^z+1)(e^{x-z}+1)} \tag{3.31}$$

we see that the required function φ has the form

$$\varphi = \frac{\alpha(\Omega)}{T^5} \frac{1}{(e^b+1)\Phi(T)}, \tag{3.32}$$

where $\alpha(\Omega)$ depends on the angle and the function $\Phi(T)$ is determined by equation (1.9).

The temperature dependence of the distribution function perturbed by the deformation potential of the sound wave may also be calculated by a somewhat different method. Returning to equation (3.29), we express the function $\varphi(\varepsilon,\ \Omega)$ as a series of Legendre polynomials and limit ourselves, as usual, to the first term of the expansion.

We thus assume that $\varphi(\varepsilon, \Omega) \approx \varphi(\varepsilon)\cos\theta = \varphi(\varepsilon)\, p_x/p$ (p is the momentum of an electron and p_x its projection on the direction of propagation of the sound wave). Equation (3.29) can now be written

$$
\begin{aligned}
f_0^2 e^{\varepsilon/T} \frac{|\xi|}{\varepsilon} \varepsilon_{ik}\dot{u}_{ik} = & \int_0^\infty N_0 |V|^2 \left(1 + \frac{\xi\xi' - \Delta^2}{\varepsilon\varepsilon'}\right) \times \\
& \times [\varphi(\varepsilon') - \varphi(\varepsilon)] \frac{p_x}{p} f_0 f_0' e^{\varepsilon'/T} \times \\
& \times \delta(\varepsilon' - \varepsilon - \hbar\omega)\, q^2 dq\, d\cos\vartheta\, d\varphi + \\
& + \int_0^{\varepsilon-\Delta} N_0 |V|^2 \left(1 + \frac{\xi\xi' - \Delta^2}{\varepsilon\varepsilon'}\right) [\varphi(\varepsilon') - \varphi(\varepsilon)] \times \\
& \times \frac{p_x}{p} f_0 f_0' e^{\varepsilon/T} \delta(\varepsilon - \varepsilon' - \hbar\omega)\, q^2 dq\, d\cos\vartheta\, d\varphi + \\
& + \int_{\varepsilon+\Delta}^\infty N_0 |V|^2 \left(1 - \frac{\xi\xi' - \Delta^2}{\varepsilon\varepsilon'}\right) [\varphi(\varepsilon') - \varphi(\varepsilon)] \times \\
& \times \frac{p_x}{p} f_0 f_0' e^{\hbar\omega/T} \delta(\varepsilon' + \varepsilon - \hbar\omega)\, q^2 dq\, d\cos\vartheta\, d\varphi + \\
& + \int_0^\infty N_0 |V|^2 \left(1 + \frac{\xi\xi' - \Delta^2}{\varepsilon\varepsilon'}\right) f_0 f_0' e^{\varepsilon'/T} \varphi(\varepsilon') \frac{q_x}{p} \times \\
& \times \delta(\varepsilon' - \varepsilon - \hbar\omega)\, q^2 dq\, d\cos\vartheta\, d\varphi + \\
& + \int_0^{\varepsilon-\Delta} N_0 |V|^2 \left(1 - \frac{\xi\xi' - \Delta^2}{\varepsilon\varepsilon'}\right) f_0 f_0' e^{\varepsilon/T} \varphi(\varepsilon') \frac{q_x}{p} \times \\
& \times \delta(\varepsilon - \varepsilon' - \hbar\omega)\, q^2 dq\, d\cos\vartheta\, d\varphi + \\
& + \int_{\varepsilon+\Delta}^\infty N_0 |V|^2 \left(1 - \frac{\xi\xi' - \Delta^2}{\varepsilon\varepsilon'}\right) f_0 f_0' e^{\hbar\omega/T} \varphi(\varepsilon') \frac{q_x}{p} \times \\
& \times \delta(\varepsilon' + \varepsilon - \hbar\omega)\, q^2 dq\, d\cos\vartheta\, d\varphi.
\end{aligned} \tag{3.33}
$$

For the momentum q_x, averaged over the azimuthal angle φ, we have (see, for example, /13/)

$$q_x = 2\pi (p_x/p)\, q \cos\vartheta$$

(ϑ is the angle between the vectors $\boldsymbol{p}$ and $\boldsymbol{q}$, and $\cos\vartheta \approx -q/2p$). We now integrate over ϑ; the rest of the calculation is similar to the derivation of (3.31). A solution is sought in the form $\varphi(\varepsilon) = \varphi_0 + \varphi_1(\varepsilon)$, where $\varphi_1 \ll \varphi_0$ ($\varphi_0 = \text{const}$). Integrating over $z = \varepsilon/T$ and the angle in the momentum space of the electrons, we get

$$
\begin{aligned}
\int_b^\infty \frac{e^z}{(e^z+1)^2} \varepsilon_{ik}\dot{u}_{ik}\, dz\, d\Omega = & |V'|^2 T^5 \int_0^\infty \frac{2x^4 dx}{e^x - 1} \int_b^\infty \frac{dz\, d\Omega \varphi_0}{(e^z+1)(e^{-z-x}+1)} + \\
& + |V'|^2 \frac{T^5}{2} \int_{2b}^\infty \frac{x^4 dx}{e^x - 1} \int_b^\infty \frac{e^z dz\, d\Omega \varphi_0}{(e^z+1)(e^{x-z}+1)}.
\end{aligned} \tag{3.34}
$$

Hence it is clear that the required function φ is characterized by the temperature dependence (3.32)

$$\varphi = \frac{\text{const}}{T^5} \frac{1}{(e^{b} + 1)\,\Phi(T)},$$

where $\Phi(T)$ is determined by equation (1.9).

Impurities are not a factor in the calculation of the viscous loss /2C, 153/. This is due to the fact that impurities cause elastic scattering of the electrons, which, as stated above, cannot lead here to establishment of an equilibrium. This may be verified formally by introducing a term describing elastic scattering by impurities in the kinetic equation (3.29). In integration over the angle, this term drops out, which confirms our assertion.

As mentioned above (see Secs. 2 and 4 of this chapter), scattering of electrons by impurities is an important factor in the presence of anisotropy of the energy gap or a multiband structure. In specimens which are not sufficiently pure, i. e., when $l \lesssim \xi_0$, impurity scattering does not change the ratio γ^s/γ^n. For example, it is pointed out in /156/ that the temperature dependence of sound attenuation in In films does not depend on their thickness.

Of course, the foregoing discussion does not apply to paramagnetic impurities. The exchange interaction of electrons with impurity atoms possessing spin strongly influences the Cooper pair condensate. A theory of superconducting alloys with paramagnetic impurities has been developed by Abrikosov and Gor'kov /199/. The influence of paramagnetic impurities on sound attenuation was investigated by the method of correlation functions in Kadanoff and Falko /200/. Sound absorption in the current state was discussed in /273/. The discussion of /273/ was generalized in /290/ to the case of paramagnetic impurities.

Neither have we taken sound attenuation due to heat conduction into consideration. In /153, 154/ it was shown that this attenuation is small in the case of a normal metal. The contribution to the energy dispersion from the temperature change caused by the sound wave is proportional to $(T/\mu_0)^2$ /154/. It is easy to see that sound attenuation due to heat conduction will also be small in this case. Note that up to now the Joule loss has been calculated only in the τ-approximation.

Once the increment to the distribution function of the electron excitations has been determined, the dissipation function $W = TS$ is calculated (S is the entropy of the electron excitation gas). This function directly determines the absorption of sound energy. The entropy is then expressed in the usual combinatorial form:

$$S = \sum_k \{(f_k - 1)\ln(1 - f_k) - f_k \ln f_k\}. \tag{3.35}$$

This expression involves the distribution function f, obtained by solving the kinetic equation (3.26). Using (3.27) we find the dissipation function to be

$$W = T\dot{S} = -T\sum_{k}\dot{f}_{k}\frac{g}{(1-f_0)f_0} \tag{3.36}$$

(g is an increment to the distribution function, $g = (\partial f_0/\partial\varepsilon)\ c(\varepsilon)$ (see above)). Substituting $\dot{f}_k$ from (3.28), we obtain

$$\begin{aligned}\dot{S} = \sum_{k}\frac{g_k}{(1-f_k)f_k}\sum_{q}N_0|V|^2\left(1+\frac{\xi\xi'-\Delta^2}{\varepsilon\varepsilon'}\right)\times \\ \times\{[c(\varepsilon')-c(\varepsilon)]\,f_0f_0'e^{\frac{\varepsilon+\hbar\omega}{T}}\delta(\varepsilon'-\varepsilon-\hbar\omega)+ \\ +[c(\varepsilon')-c(\varepsilon)]\,f_0f_0'e^{\frac{\varepsilon}{T}}\delta(\varepsilon-\varepsilon'-\hbar\omega)\}- \\ -\sum_{k}\frac{g_k}{(1-f_k)f_k}\sum_{q}N_0|V^2|\frac{1}{2}\left(1-\frac{\xi\xi'-\Delta^2}{\varepsilon\varepsilon'}\right)\times \\ \times[c(\varepsilon')+c(\varepsilon)]\,f_0f_0'e^{\frac{\hbar\omega}{T}}\delta(\varepsilon'+\varepsilon-\hbar\omega).\end{aligned} \tag{3.37}$$

We integrate over the angle in the phonon momentum space. Next we change over from integration over electron momentum to integration over ξ, then to integration over the energy of the electron excitations ε. Using the approximation (1.13), we find

$$\begin{aligned}\dot{S} = \int_0^\infty\frac{4x^2dx}{e^x-1}\int_b^\infty[\varphi(z+x,\ \Omega')-\varphi(z,\Omega)]^2\times \\ \times\frac{e^{z+x}dz}{(e^z+1)(e^{z+x}+1)}+\int_{2b}^\infty\frac{x^2dx}{e^x-1}\int_b^{x-b}[\varphi(z,\Omega)- \\ -\varphi(x-z,\ \Omega')]^2\frac{e^x dz}{(e^z+1)(e^{z-x}+1)}\ .\end{aligned} \tag{3.38}$$

We again expand $\varphi(z,\Omega')$ in powers of $\boldsymbol{q}'=\boldsymbol{p}-\boldsymbol{p}^*$, after which we substitute the expression (3.32) obtained above for the function φ. Hence the following final expression for the dissipation function:

$$\dot{W} = \frac{\text{const}}{T^5}\frac{1}{(e^b+1)^2\Phi(T)} \tag{3.39}$$

($\Phi(T)$ is determined by equation (1.9)). The dissipation function for a normal metal is obtained by setting $b = \Delta/T = 0$.

Thus, the absorption coefficient of long wavelength sound in superconductors can be given by the following formula:

$$\gamma^s = \gamma^n\frac{4\Phi(T_c)}{(e^b+1)^2\Phi(T)}, \tag{3.40}$$

where $\gamma^n = \text{const}\cdot T^{-5}$ is the sound absorption coefficient in a normal metal, an expression for which was obtained by Akhiezer /153/, and

$$\Phi(T) \approx 96\zeta(4)\ln(1+e^{-b}) + \\ + 40\sum_{s=1}^{\infty} s^{-5}e^{-2bs}(2b^4s^4+4b^3s^3+6b^2s^2+6bs+3) - \\ -16\ln(e^b+1)\sum_{s=1}^{\infty} s^{-4}e^{-2bs}(4b^3s^3+6b^2s^2+6bs+3), \quad (3.41)$$

$$\zeta(s) = \sum_{n=1}^{\infty} n^{-s}.$$

Thus, one sees that the absorption coefficient decreases with decreasing temperature as described by (3.40). The ratio γ^s/γ^n is a universal function of temperature, as in the case of ultrasonic attenuation (see Sec. 1). The temperature dependence of the attenuation is caused by the decrease in the number of electron excitations as $T \to 0$.

11. SOUND ABSORPTION BY PHONONS

Sound energy may also be absorbed by phonons. This problem has been solved /20/ in principle by the method used in the preceding section for the calculation of the absorption coefficient of long wavelength sound by electron excitations.

The sound field changes the frequency of a phonon

$$\omega' = \omega + \alpha_{i\nu} u_{i\nu};$$

here $\alpha_{i\nu}$ is a tensor which depends on the direction of the phonon wave vector and $u_{i\nu}$ is the deformation tensor. The kinetic equation for the distribution function of phonons interacting with electrons under the influence of the sound field on the superconductor is

$$- N_0^2 e^{\hbar\omega/T}\frac{\hbar\omega}{T}\alpha_{i\nu}\dot{u}_{i\nu} = \int |V|^2\{(u_k u_{k'} - \\ - v_k v_{k'})^2[f'(1-f)(N+1) - Nf(1-f')]\times \\ \times\delta(\varepsilon'-\varepsilon-\hbar\omega) + (u_k v_{k'} + u_{k'}v_k)^2\times \\ \times[(N+1)ff' - N(1-f)(1-f')]\}\times \\ \times\delta(\varepsilon'+\varepsilon-\hbar\omega)\,p^2dp\,d\cos\vartheta\,d\varphi. \quad (3.42)$$

The perturbation of the phonon distribution function $R = N - N_0$ can be found from this equation by a method similar to that employed in Chapter I, Sec. 4, for calculation of the lattice thermal conductivity. We seek a solution in the form $R = -N_0^2 e^x r(x)/T$. The electronic distribution function (3.27) is substituted into the integral. As a result, we find

$$r(x) = \text{const}\, x \left[\int_b^\infty \frac{e^z dz}{(e^z+1)(e^{z+x}+1)} + D(x) \int_b^{x-b} \frac{dz}{(e^z+1)(e^{x-z}+1)} \right]^{-1}, \tag{3.43}$$

$$D(x) = \begin{cases} 0 \text{ for } x < 2b, \\ 1 \text{ for } x \geqslant 2b. \end{cases}$$

Next, writing the expression for the entropy of the phonon gas in combinatorial form and calculating $\dot{S}$, we find the dissipation function

$$W_p = -T \sum_q R\dot{N}/N_0(N_0+1).$$

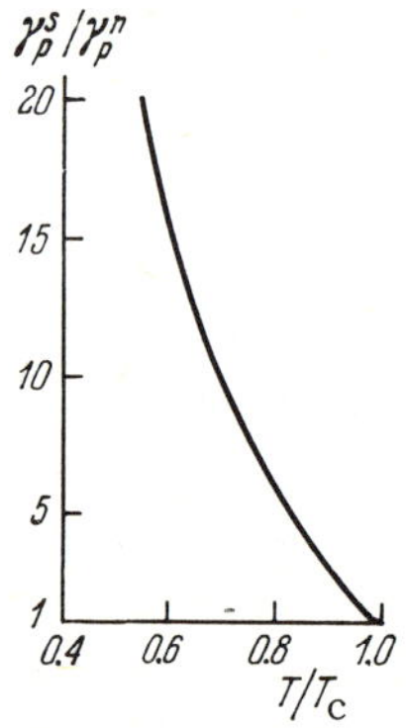

FIGURE 25. Sound absorption by phonons.

Evaluation leads to the result (Figure 25)

$$\frac{\gamma_p^s}{\gamma_p^n} = \frac{F(T)}{F(T_c)}, \tag{3.44}$$

$$F(T) = -8(b^4+b^3)(e^b-1)^{-1} + 6\zeta(3)(e^b+1) -$$

$$- 3(e^b + 1)\sum_{s=1}^{\infty} s^{-3}e^{-2bs}(4b^2s^2 + 4bs + 2) +$$
$$+ 6\zeta(4)(e^b - 1) - (e^b - 1)\sum_{s=1}^{\infty} s^{-4}e^{-2bs}(8b^3s^3 +$$
$$+ 12b^2s^2 + 12bs + 6) + 32b^3(e^{2b} - 1)^{-1} -$$
$$- a^4\sum_{s=1}^{\infty}\{se^{-2bs}\,\mathrm{Ei}\,[-s(2b - a)]\} + 6\sum_{s=1}^{\infty} s^{-3}e^{-2bs};$$
$$a \approx 2b - 0{,}16,\ \zeta(s) = \sum_{n=1}^{\infty} n^{-s},\quad a \to 2b \text{ for } T \to T_c.$$

Here γ_d^s and γ_p^n are the coefficients of sound absorption by phonons in the superconducting and normal state (γ_p^s/γ_p^n is also a universal function of temperature). The total absorption coefficient is

$$\gamma^s = \gamma_e^s + \gamma_p^s.$$

The absorption of sound by phonons increases with decreasing temperature. This is due to an increase in the free path of the phonons, as a consequence of the decrease in the number of electron excitations. This absorption mechanism is significant in the intermediate temperature range.

With further decrease in temperature, the phonons begin to be scattered mainly by the impurities, lattice defects, and crystal boundaries, rather than by the electron excitations, whose number decreases exponentially. When this happens the attenuation begins to decrease with decreasing temperature, because of the power-function decrease in the number of thermal phonons. The pattern becomes equivalent to low temperature sound absorption in dielectrics ($\gamma^s \to 0$ as $T \to 0$). In the intermediate temperature range ($T/T_c \approx 0.3$), therefore, the quantity γ^s passes through a maximum similar to the maximum of the lattice thermal conductivity (see Chapter I, Sec. 4).

12. SOUND ABSORPTION IN THE INTERMEDIATE STATE

Sound absorption in the intermediate state is characterized by a host of important features. The first factor to be mentioned is an additional mechanism of sound wave absorption, investigated by Andreev and Bruk in /203/, associated with the system of alternating normal and superconducting layers characteristic for the intermediate state.

The critical magnetic field depends, of course, on the specimen's pressure and temperature. Its magnitude therefore changes in the presence of a sound wave. The magnetic field in the normal layers is equal to the critical field, and therefore a change in the latter results in a shift of the interphase boundary. A variable magnetic field and Foucault currents then arise in the normal layers. The Joule heat thus evolved represents an additional mechanism for dissipation of sound wave energy.

In this case the critical field H_c can be written

$$H_c = H_{c0}(1 + \alpha u_{ii}) + \frac{dH_c}{dT} T'. \tag{3.45}$$

Here H_{c0} is the critical field in the absence of sound; $\alpha \approx 1$; $u_{ii} = \operatorname{div} \boldsymbol{u}$ ($\boldsymbol{u}$ is the lattice displacement in the sound field, $\boldsymbol{u} = \boldsymbol{u}_0 \exp[i(\boldsymbol{kr} - \omega t)]$), and T' is the temperature change caused by the emission or absorption of heat due to the motion of the interphase boundaries.

The quantity α can be expressed in terms of the derivatives $(\partial H_c/\partial T)_p$ and $(\partial H_c/\partial p)_T$. Henceforth we assume that $\lambda \gg \delta$, $\lambda \gg a^n$ (λ is the wavelength, δ the skin depth, a^n the thickness of the normal layers).

The full system of equations for the magnetic field has the form (see, for example, /204/, Sec. 45)

$$\frac{\partial \boldsymbol{H}}{\partial t} - \frac{c^2}{4\pi\sigma} \Delta \boldsymbol{H} = \operatorname{rot} [\boldsymbol{vH}], \tag{3.46}$$

$$\operatorname{div} \boldsymbol{H} = 0. \tag{3.46'}$$

The boundary conditions are

$$\begin{gathered} E_{x|z=\pm a^n/2} = 0, \quad E_y + \frac{\dot{U}}{c} H_{c0|z=\pm a^n/2} = 0, \\ H_x - H_{c|z=\pm a^n/2} = 0, \\ H_z - H_{c0} \frac{\partial U}{\partial x}_{|z=\pm a^n/2} = 0. \end{gathered} \tag{3.47}$$

Here

$$\boldsymbol{E} = \frac{c}{4\pi\sigma} \operatorname{rot} \boldsymbol{H} - \frac{1}{c} [\boldsymbol{vH}]$$

is the electric field, U the deviation of the interphase boundary from its equilibrium position ($U \approx u = |\boldsymbol{u}|$), $\boldsymbol{v} = \dot{\boldsymbol{u}}$ is the velocity of the medium, and $\boldsymbol{H}_{c0}$ is the magnetic field in the absence of sound.

The coordinate system is chosen as in Chapter I, Sec. 10, so that the z-axis is perpendicular to the interphase boundary and the xy-plane lies in the middle of the normal layer. In addition, the x-axis is parallel to the direction of $\boldsymbol{H}_{c0}$.

The full system of equations contains (3.46) and (3.47) plus the heat balance equation:

$$ac_V\frac{\partial T'}{\partial t} = Q(\dot{\xi}_- - \dot{\xi}_+), \tag{3.48}$$

$$Q = -\frac{T}{4\pi} H_c \frac{\partial H_c}{\partial t},$$

which determines the quantity T' appearing in (3.45). In (3.48), $a = a^s + a^n$ (a^s is the thickness of the superconducting layer); c_V is the specific heat; $\xi_\pm = \xi_{z=\pm a^n/2}$ is the displacement of the interphase boundary.

The time average of the energy dissipation per unit volume is

$$W = \frac{1}{a}\int_{-a^n/2}^{a^n/2} dz \frac{|\boldsymbol{j}|^2}{2\sigma},$$

where $\boldsymbol{j}$ is the current density,

$$\boldsymbol{j} = \frac{c}{4\pi} H_{c0}\frac{\partial h}{\partial z}.$$

For the value of $\boldsymbol{h}$ determined by the equality

$$\boldsymbol{H} - \boldsymbol{H}_{c0} = \frac{i}{\omega}\,\mathrm{rot}\,[\boldsymbol{v}\boldsymbol{H}_{c0}] + H_{c0}\boldsymbol{h},$$

only the x-component is different from zero. A system of equations for $h = h_x$ can be obtained from (3.46) and (3.47). The solution has the form

$$h = \frac{u_{ii}[\alpha + 1 - (\boldsymbol{mn})^2]\cos qz}{\cos(q a^n/2) - iq\delta D\sin(qa^n/2)}, \tag{3.49}$$

$$D = \frac{T\delta}{4\pi ca}\left(\frac{\partial H_c}{\partial T}\right)^2.$$

The final expression for the sound absorption coefficient (i. e., the ratio of W to the energy flow in the sound wave) is

$$\gamma = \left(\frac{cH_c^2}{8\pi}\right)\frac{2[\alpha + 1 - (\boldsymbol{mn})^2]^2[\mathrm{sh}(a^n/\delta) - \sin(a^n/\delta)]}{\rho\sigma(a^s + a^n)\delta s_l^3} \times$$
$$\times\left\{\cos^2\left(\frac{a^n}{2\delta}\right) + \mathrm{sh}^2\left(\frac{a^n}{2\delta}\right) + D[\mathrm{sh}(a^n/\delta) + \sin(a^n/\delta)] + \right.$$
$$\left. + 2D^2\left[\sin^2\left(\frac{a^n}{2\delta}\right) + \mathrm{sh}^2\left(\frac{a^n}{2\delta}\right)\right]\right\}^{-1}. \tag{3.50}$$

In the limiting case of very low frequencies, when $a^n \ll \delta$, the absorption at $T \ll T_c$ is described by the equation

$$\gamma = \frac{(a^n)^3 [\alpha + 1 - (\boldsymbol{mn})^2]^2}{24 (a^s + a^n) \rho \sigma u_l^3 \delta^4} \left(\frac{cH_c}{2\pi} \right)^2 \tag{3.51}$$

($\boldsymbol{m}, \boldsymbol{n}$ are unit vectors in the directions of $\boldsymbol{H}_{c0}$ and $\boldsymbol{k}$; ρ is the density of the metal; a^s is the thickness of the superconducting layers; u_l is the velocity of longitudinal sound; σ is the steady-state conductivity; it is assumed that the free path of the electrons is significantly smaller than a^n and δ, so that it is legitimate to use σ). One sees that here the absorption is proportional to the squared sound frequency ω^2 and is practically independent of the temperature. As T increases, however, the temperature dependence becomes substantial.

In fact, if $T \approx T_c$ the absorption coefficient is

$$\gamma = \frac{(a^n)^3 [\alpha + 1 - (\boldsymbol{mn})^2]^2}{24 (a^s + a^n) \rho \sigma u_l^3 \delta^4} \left(\frac{cH_c}{2\pi} \right)^2 \left\{ 1 + D \frac{a^n}{\delta} \right\}^{-2}. \tag{3.52}$$

The function D figuring in (3.49) now determines the temperature dependence of the absorption.

In the opposite limiting case, $a^n \gg \delta$, the absorption

$$\gamma = \frac{4 [\alpha + 1 - (\boldsymbol{mn})^2]^2}{(a^n + a^s) \rho \sigma u_l^3 \delta} \left(\frac{cH_c}{8\pi} \right)^2 \tag{3.53}$$

is proportional to $\omega^{1/2}$.

The ratio of the sound absorption coefficient associated with the mechanism under discussion to the usual electronic absorption is a quantity of the order of $(e^2/c)\,(v_F/c)\,(a/\xi_0)^2$ $(a = a^n + a^s)$. This ratio may be greater than unity, because of the factor a/ξ_0. In that case the mechanism of low-frequency sound absorption associated with the movement of the phase boundary begins to be a major factor. The frequency dependence of the absorption in this case is very peculiar. In the very low frequency range the attenuation is proportional to ω^2; then, with increasing frequency, a region appears (when $a^n \gg \delta$) where the absorption is proportinal to $\omega^{1/2}$. At even higher frequencies, where absorption is due to the usual mechanism /154/, one again has $\gamma \sim \omega^2$. The excellent agreement of the results with equation (3.53) has been noted in /205/, which contains an experimental investigation of the absorption of sound at frequencies varying in the interval from 21 to 90 mHz. The sound wave was absorbed in pure superconducting lead in the intermediate state.

We have been discussing the absorption of low-frequency sound. Andreev /206/ has investigated the attenuation of high-frequency sound vibrations by superconductors in the intermediate state. The wavelength of the sound was of the same order of magnitude as the thickness of the layers. It was also assumed that $a^n \ll l$ and $a^n \ll R$ (R is the Larmor radius in the critical magnetic field).

It turns out that in this case one can determine the period of the laminar structure of the superconductor by means of ultrasonic measurements. A knowledge of the period is highly important, since it enables us to determine the value of the surface tension coefficient for the boundary between the normal and superconducting phases. This coefficient is a single-valued function of the period of the structure /207, 208/, $a(H)=a^n(H)/\eta\,(H)$ (see also /204/, Sec. 44).

The treatment was carried out at temperatures such that $T \ll T_c$. Under these conditions ultrasonic absorption in the superconducting phase is extremely low in view of the exponentially small number of electron excitations. The absorption is determined mainly by the normal phase. Furthermore, the low temperature ensures that the above-mentioned conditions $a^n \ll l$, $a^n \ll R$, under which the calculation of /206/ is valid, are satisfied.

The following situation is very convenient from the standpoint of experimental setup. The sound propagates along the axis of a cylindrical specimen (below the x-axis is directed along the sound wave vector), while the field, directed along the z-axis (the planes $x=0$ and $x=a$ coincide with the boundaries of the normal layer), is perpendicular to the axis of the cylinder.

The electron energy is modified in the sound wave field and becomes /153, 154/ (see Sec. 8)

$$\varepsilon = \varepsilon_0(\boldsymbol{p}) + \varepsilon_{ik}(\boldsymbol{p})\, u_{ik}$$

($\varepsilon_0(\boldsymbol{p})$ is the energy of the electron in the absence of sound, u_{ik} is the deformation tensor). The first step in solving the problem is to determine the electronic distribution function perturbed by the sound field, which, as usual, has the form

$$f = f_0 + \chi \partial f_0/\partial\varepsilon.$$

If $R \gg l$ one can disregard the effect of the magnetic field on the electrons. One can then write the kinetic equation in the range $0 \ll x \ll a$ as

$$v_x \partial\chi/dx + \nu\chi = ev_i E_i - \Lambda_{ik}\dot{u}_{ik}. \tag{3.54}$$

Here ν is the collision time, $\Lambda_{ik} = \lambda_{ik} - \overline{\lambda_{ik}}$ (the averaging is carried out over the Fermi surface). The first term on the right side of (3.54) describes the effect on the electrons of the electric field induced when the sound propagates in the metal. The deformation and electric field give rise to a current whose density is related to the transverse electric field by Maxwell's equation

$$\operatorname{rot}\operatorname{rot}\boldsymbol{E} = \frac{4\pi i\omega}{c^2}\,\boldsymbol{j}$$

(ω is the frequency of the sound).

We seek a solution for equation (3.54) in the form $\chi = \varphi + \psi$, where ψ and φ are respectively solutions of the equations obtained from (3.54) when only the first and the second term appear in the right-hand side.

The time average of the energy of a sound wave absorbed in unit volume is

$$\bar{Q} = \frac{1}{a^n + a^s}\int\limits_0^{a^n} Q(x)\,dx, \quad Q(x) = \frac{1}{(2\pi)^3}\int \frac{ds}{v}\,\nu\,|\chi|^2 \tag{3.55}$$

(ds is an element of the Fermi surface).

The expression for $\bar{Q}$ becomes

$$\bar{Q} = \frac{\eta}{(2\pi)^3}\int \frac{\nu\,ds}{v}\sum_n \{|\varphi_n|^2 + |\psi_n|^2\}, \tag{3.56}$$

where $\eta = a^n/(a^n + a^s)$ is the concentration of the normal phase, and φ_n and ψ_n are the Fourier components of the terms of the function χ; for example,

$$\varphi_n = \frac{1}{2a^n}\int\limits_{-a^n}^{a^n} \varphi(x)\,e^{-iq_n x}\,dx.$$

Substituting the functions φ_n and ψ_n found by solving (3.54) (with boundary conditions $\chi(v) + \chi(-v) = 0$ at $\chi = 0, a$) into (3.56), we find the ratio of the absorption coefficients in the superconducting and normal phases:

$$\gamma^s/\gamma^n = \eta\Phi(ka^n/\pi), \tag{3.57}$$

where

$$\Phi(x) = \frac{1}{\pi^2}\left\{\pi^2 - \frac{1}{x^2} - 2\psi'(1+x) - \right.$$
$$\left. - \cos\frac{\pi x}{2}\left[\psi'\left(\frac{1+x}{2}\right) - \psi'\left(1 + \frac{x}{2}\right) - \frac{2}{x^2}\right]\right\}$$

(ψ' is the second derivative of the logarithm of the Γ-function).

In the limiting case, when the wavelength satisfies the condition $\lambda \ll a^n$, we obtain from (3.57)

$$\gamma^s/\gamma^n = \eta. \qquad (3.58).$$

At the other extreme, $\lambda \gg a^n$ we have $\Phi(x) = 7\zeta(3)x/\pi^2$, which gives

$$\gamma^s/\gamma^n = (7\zeta(3)/\pi^3)\,\eta k a^n. \qquad (3.59).$$

The absorption is then of the same order of magnitude as that of absorption in a normal metal /154/, in which the free path equals a^n.

From (3.56), (3.58), and (3.59) it is clear that the absorption enables us actually to determine the thickness of the normal layer a^n, and then, via the relationship $\eta = 2(H/H_c) - 1$, the period of the structure, $a = a^n + a^s$, which is $a(H) = a^n(H)/\eta(H)$.

We have considered the limiting case $R \gg l$. In the case $R \ll l$, the magnetic field must be incorporated in the kinetic equation and we obtain

$$v_x \frac{\partial \chi}{\partial x} + \Omega \frac{\partial \chi}{\partial \tau} + \nu \chi = e v_i E_i - \Lambda_{ik} \dot{u}_{ik}$$

(Ω is the cyclotron frequency in the field H_c, τ is the dimensionless period of an electron in the magnetic field). Solution of this equation, which is largely similar to solution of the kinetic equation for a normal metal /209/, leads to the following expression for the absorption coefficient:

$$\gamma = \gamma_0 + \Delta\gamma,$$

where γ_0 is a monotonic function of the external magnetic field:

$$\gamma_0 = \eta \Phi(ka/\pi)\, \gamma_0^n(H_c)$$

($\gamma_0^n(H_c)$ is the monotonic part of the absorption in a normal metal), and $\Delta\gamma$ is an oscillating function:

$$\Delta\gamma \sim \gamma_0^n(H_c)\,\eta k a^n \left(\frac{a^n}{R}\right) \times$$
$$\times \sum_{m=1}^{\infty} \frac{m^{1/2}\,[1 - (-1)^m \cos k a^n]}{(m^2 - k^2 a^{n}/\pi^2)^2} \sin\left(\frac{\pi R_{ext}}{a^n} m \pm \frac{\pi}{4}\right) \qquad (3.60)$$

($R_{ext} = c\,(p_y^{(2)} - p_y^{(1)})_0/eH_c$; the expression in parentheses is evaluated at the extremum value of p_z). Measurement of both the monotonic and the oscillating part of the absorption enables us in this case to determine the period of the structure in the intermediate state.

Experimental observation of oscillatory ultrasonic absorption in the intermediate state /206/ is quite difficult. It is possible only when

$$l \gg D_{ext} \gg a^n$$

(l is the free path, D_{ext} the extremum diameter of the orbit). This new resonance phenomenon was first observed recently by Shepelev, Ledenev, and Filimonov /310/. A single crystal of extremely pure helium was used. The authors overcame the substantial experimental difficulties and observed the oscillatory behavior of $\gamma(H)$.

13. ABSORPTION OF TRANSVERSE SOUND WAVES

The attenuation of transverse waves in superconductors has not been investigated nearly as extensively as the absorption of longitudinal sound. The singular features of transverse sound attenuation have been pointed out in /210, 211/ with reference to Sn and In, and in /212–214/, where the corresponding properties of Al were discussed.

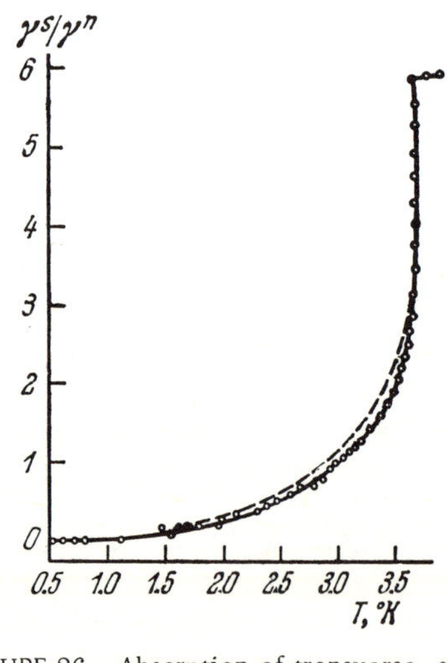

FIGURE 26. Absorption of transverse sound.

There are two regions, sharply distinguished by their temperature dependence. In the first, very close to T_c, a very abrupt drop in absorption is observed (Figure 26). Thus, for example, in /210/, when the temperature fell only 0.01° below T_c, absorption of transverse sound of frequency $\omega \sim 30$ mHz dropped to half its original value. After this sharp drop the temperature dependence is adequately described by the usual equation (3.3) of the BCS theory.

In /5, 215–217/ it was conjectured that this sharp drop is due to the field associated with the transverse sound wave being screened by the superconducting currents.

However, further research has shown that the effect is not universal. Thus, it was shown in /218/ that attenuation of transverse sound in Va and Ta exhibits no anomalies, and its temperature dependence in the entire temperature range is exactly like that characterizing the absorption of longitudinal waves. An investigation of pure Ga ($ql > 100$) /219/ showed that several directions of sound propagation are associated with a range of sharp drops in absorption, while for other directions no such anomalous drop is observed.

References /202, 215, 220, 221/ discuss the role of electric fields in sound absorption. In /221/ it was shown that certain peculiar features of absorption should be observed in various cases. For example, the temperature dependence of absorption due to fields in Pippard superconductors with parameters satisfying the inequalities $\omega \ll T \ll \Delta \ll qv$ is described by the equation

$$\gamma \sim \omega \frac{\Delta}{T} \exp\left(-\frac{\Delta}{T}\right)\left(a\omega^2 + \frac{\Delta}{\omega}\right)^{-2} \qquad (3.61)$$

(the constant a is independent of the temperature). In /221/ it was also shown that the contribution of the fields to sound absorption is usually small (compared with the usual deformation mechanisms).

In /202/ the absorption coefficient was calculated, allowing for both the deformation potential and the contribution of macroscopic fields. The full system of equations was investigated, comprising the equations of elasticity theory, Maxwell's equations, and the kinetic equation for electron excitations. It was shown that at frequencies $\omega \lesssim 10^8$ sec^{-1} the fields are not a factor and that absorption is associated as usual with a direct deformation interaction. The authors of /202/ demonstrated that the contribution of the fields becomes significant only at frequencies 10^9–10^{10} sec^{-1}.

It should be noted that at such high frequencies the sound wave period usually satisfies the condition $T \ll \tau$ (τ is the relaxation time), and sound absorption can be viewed as a process involving direct absorption of a sound quantum by electron excitations (see Sec. 1). It is not accidental that experimentally observed sound absorption at frequencies (say $\omega \sim 10^8$ sec^{-1}) less by at least two orders of magnitude than the value cited in /202/ is described quite well by equation (3.3). Thus, consideration of the attenuation of transverse sound in superconductors leads to the conclusion that in the main frequency range, the major role is played by absorption associated with direct deformation interaction.

The authors of /222, 223/ observed a frequency dependence of the ratio γ^s/γ^n for transverse sound propagating in Al and Sn. As a matter of fact, the calculations in /202/ imply a frequency dependence. This effect has been investigated for longitudinal sound in Sn /223/, Hg /179/, Pb /177/ and Nb /311/.

At first sight, this dependence seems surprising, since, as shown previously with regard to attenuation of longitudinal sound in superconductors, the ratio γ^s/γ^n is described by a universal function of the temperature, which is independent of the frequency. That a frequency dependence can appear is due to the marked difference between the free paths l^s_{ep} and l^n_{ep}, which describe the scattering of electrons by phonons in the superconducting and normal states, respectively. The free paths l^s_{ed} and l^n_{ed} are identical (see Chapter I, Sec. 2). The length of the free path l^s_{ep}, however, decreases significantly with decreasing temperature and may differ considerably from the value of l^n_{ep} corresponding to a normal metal at the same temperature. The relation between l and λ is important for the frequency dependence (see for example, /109/). A situation is possible in which $l^s_{ep} \gg \lambda$ and $l^n_{ep} \ll \lambda$. In that case the ratio γ^s/γ^n depends on the frequency at a given temperature, and this can be observed experimentally. Naturally, the magnitude of this effect depends on the degree of purity of the specimen, since it is absent if the electrons are scattered by impurities; it is due only to electron-phonon interactions. In /311/ it is observed that the frequency dependence becomes more marked with a decrease in the impurity concentration in superconducting Nb. This agrees fully with Andreev's explanation, which was also proposed independently by the authors of /311/.

Reference /224/, an experimental study of the absorption of transverse ultrasound in a magnetic field, should also be mentioned. The measurements were carried out at frequency $\omega \sim 7.5 \cdot 10^7\ \mathrm{sec}^{-1}$. An absorption peak was detected in the near-critical temperature range. The position of the peak was observed to depend on the magnitude of the magnetic field and the polarization of the sound wave.

14. PECULIARITIES OF KINETIC PHENOMENA IN TYPE-II SUPERCONDUCTORS

Type-II superconductors, which are in the mixed state, are characterized by the presence of a system of vortex lines /225/.

The superconducting ordering parameter $\Delta(\boldsymbol{r})$ approaches zero on the axis of the vortex. The vortical structure and the normal component associated with its appearance are additional factors to be considered in investigation of kinetic phenomena in type-II superconductors. The experimentally observed peculiarities of the processes of heat transfer and sound absorption in the mixed state are due to the effect of these factors.

The behavior of electronic thermal conductivity was investigated theoretically in /226, 227/: $\varkappa_e$ varies linearly near H_{c2} ; $\varkappa_e \sim (H - H_{c2})$. Doped specimens of lead and indium were investigated in /228/; in agreement with theory, it was observed that under conditions in which the main contribution to the heat flux is electronic, the ratio of the derivatives $d\varkappa/dH$ and dM/dH (M is the magnetization) is a universal function of the temperature.

A monotonic increase in thermal conductivity as a function of the field in the interval from H_{c1} to H_{c2} was observed in /229/, where single-crystal specimens of unordered alloys of tantallum and niobium were investigated. The measurements were carried out at a constant temperature near T_c. The main contribution to the heat flux is then due to the electronic component. Similar behavior has been noted in pure Nb /230, 231/. Reference /233/ observes a significant anisotropy of $\varkappa_e$ as a function of the angle between the direction of the heat flux and that of the vortex lines. The behavior of $\varkappa_e$ is used to determine H_{c1} and H_{c2}.

The most significant anomalies are observed in investigation of the phonon thermal conductivity of type-II superconductors. The appearance of large-scale formations at $H > H_{c1}$, which are Abrikosov vortices, results in a marked decrease in the free path of thermal phonons and a consequent sharp drop in the value of $\varkappa_p$. This drop in thermal conductivity when the field reaches H_{c1} has been observed /229/ in the low temperature range, where, as noted previously (See Chapter I, Sec. 5), phonon thermal conductivity is the major factor. Its decrease with an increase in the field implies that the electronic thermal conductivity $\varkappa_e$ in the superconductor begins to be the major factor again, since $\varkappa_e$ increases with an increase in H. It is clear from the above properties that the thermal conductivity $\varkappa(H)$ of type-II superconductors, as a function of the magnetic field in the interval $H_{c1} < H < H_{c2}$, is characterized by a minimum in the low-temperature range. This minimum has in fact been observed /229—232/. In /229/ these features were also established in the temperature dependence of $\varkappa$ (Figure 27).

In /234/ the ultrasonic absorption coefficient was calculated in a dirty superconductor alloy ($\tau T_c \ll 1$, τ is the relaxation time). It was assumed that the field was not too great ($H_{c1} < H \ll H_{c2}$).

The author determined the work performed by the electron gas of a sound wave whose length satisfies the conditions $ql \ll q\delta_0 \ll 1$, $qd \ll 1$, where d is the distance between the vortex lines. The energy dissipation is described in the hydrodynamic approximation by the following equation:

$$\overline{\langle \dot{H} \rangle}^t = \int dV \, [\overline{\eta_{iklm} \dot{u}_{ik} \dot{u}_{lm}^t} + \overline{\sigma_{ik} E_i E_k^t}] \tag{3.62}$$

(E_i is the effective electric field in the coordinate system associated with the lattice; σ_{ik} is the conductivity tensor; η_{iklm} is the viscosity coefficient tensor; u_{ik} is the deformation tensor; the brackets $\langle \ldots \rangle$ denote thermodynamic average, the bar — denotes time average).

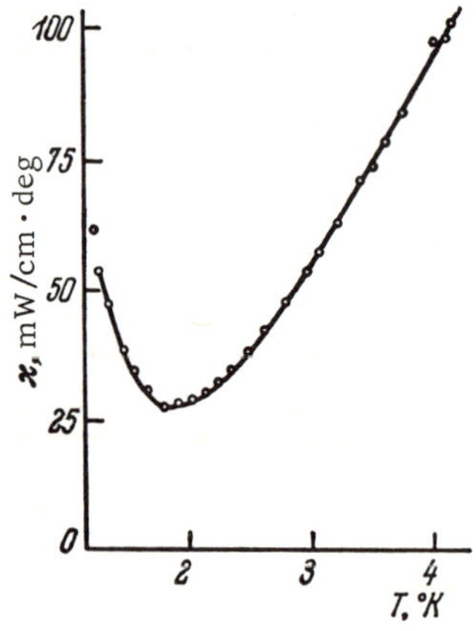

FIGURE 27. Thermal conductivity in the mixed state /229/

The first term on the right of (3.62) represents the usual absorption of a sound wave by the electron system. The associated attenuation coefficient can be expressed as

$$\gamma^s/\gamma^n = (\gamma^s/\gamma^n)_{\mathrm{BCS}} + K(T)\, B/H_{c2}$$

(B is the magnetic induction in the superconductor; the function $K(T)$ is expressed in a certain way in terms of $\varkappa_{\mathrm{GL}}$ and Δ/T, $\varkappa_{\mathrm{GL}}$ is the parameter of the Ginzburg-Landau theory: at $T=0$ we have $K(0) \approx 1$).

The principal feature of sound absorption in the mixed state stems from the second term in (3.62), which describes the ohmic loss associated with the appearance of induced electric fields. When the crystal lattice is displaced because of the passage of a sound wave, the vortex lines are entrained by the ion system of the metal.

However, the entrainment is not complete, owing to the elastic interaction between the vortices.

Induced electric fields appear because of the motion of the vortex relative to the ion lattice. When this occurs, the absorption of both longitudinal and transverse sound is strongly anisotropic. Thus, for example, the absorption coefficient for longitudinal sound is proportional to $\sin^6 \theta$, where θ is the angle between the vectors $\boldsymbol{q}$ and $\boldsymbol{H}_0$. When the sound travels in the direction of the field, the vortex lattice is not deformed and the corresponding absorption coefficient is equal to zero. Waves traveling perpendicular to the field direction are the most strongly absorbed.

In /236/ ultrasonic absorption in the mixed state in the presence of paramagnetic impurities is discussed.

References /237–241/ are devoted to a theoretical and experimental study of sound attenuation in pure type-II superconductors. In /239/ a linear dependence of $[1 - (\gamma_l^s/\gamma_l^n)]$ on $[H_{c2}(T) - H]^{1/2}$, agreeing with theory /237/, was experimentally obtained (H is the external field, γ_l the absorption coefficient for longitudinal sound in pure Nb).

The vortical structure of type-II superconductors gives rise to a phenomenon similar to the effect of ordinary magnetic resistance in a normal metal. Thermomagnetic and galvanomagnetic effects also occur. Among these are the Ettingshausen effect /240–244/ (appearance of a temperature gradient perpendicular to the passing current, where $\Delta T \sim (V_l/\varkappa)(T/T_c)^2$ /241/, V_l being the longitudinal electric voltage and $\varkappa$ the thermal conductivity), the Nernst longitudinal effect /240, 242, 245/, and the Hall effect. Emission of Peltier heat has been observed /241, 243, 244/ on the boundary separating the mixed ($H > H_{c1}$) and ordinary superconducting states.

15. NUCLEAR SPIN RELAXATION

The rate of nuclear spin relaxation can easily be calculated using the (u, v)-transformation /1, 5, 249/. Let $V_{kk'}$ denote the matrix element of the interaction of electrons with the nuclear spin, which leads to transition of an electron from a state with momentum $\boldsymbol{k}$ and spin σ to the state with momentum $\boldsymbol{k}'$ and spin σ', is denoted by $V_{kk'}$; then the perturbation operator has the form

$$H' = \sum_{k,\, k',\, \sigma} V_{kk'} a^+_{k'\sigma} a_{k\sigma}$$

($\sigma = +, -$). Transforming from the operators $a_{k\sigma}, a^{+}_{k\sigma}$ to the operators $\alpha_{k\sigma}, \alpha^{+}_{k\sigma}$ /2, 248/ via the formulas

$$a_{k+} = u_k\alpha_{k+} + v_k\alpha^{+}_{-k-}, \qquad a_{k-} = u_k\alpha_{k-} - v_k\alpha^{+}_{-k+},$$

we obtain

$$\begin{aligned} H' = \sum_{k,\,k',\,\sigma} V_{kk'} [u_k u_{k'} (\alpha^{+}_{k+}\alpha_{k'+} + \alpha^{+}_{k-}\alpha_{k'-}) + \\ + v_k v_{k'} (\alpha_{-k-}\alpha^{+}_{-k'-} + \alpha_{-k+}\alpha^{+}_{-k'+}) + u_k v_{k'} (\alpha^{+}_{k+}\alpha^{+}_{-k'-} - \alpha^{+}_{k-}\alpha^{+}_{-k'+}) + \\ + v_k u_{k'} (\alpha_{-k-}\alpha_{k'+} - \alpha_{-k+}\alpha_{k'-})] = \\ = \sum_{k,\,k',\,\sigma} V_{kk'} [(u_k u_{k'} + v_k v_{k'}) (\alpha^{+}_{k+}\alpha_{k'+} + \alpha^{+}_{k-}\alpha_{k'-}) + \\ + (u_k v_{k'} - u_{k'} v_k) (\alpha^{+}_{k+}\alpha^{+}_{k'-} + \alpha_{k+}\alpha_{k'-}) + 2v_k^2\delta_{kk'}]. \end{aligned} \tag{3.63}$$

This equation is derived using the relationship $V_{k,\,k'} = - V_{-k',\,-k}$, which is characteristic for the spin interaction. In interaction of electrons with impurities, phonons, and with the nuclear quadrupole moment, $V_{k,\,k'} = V_{-k',\,-k}$. The transition probability is determined by the usual equation $w = 2\pi |H'|^2_{ab}\rho_E$ ($\hbar = 1$, ρ_E is the statistical weight). The total probability is equal to the difference between the probabilities of absorption and induced emission:

$$\begin{aligned} W \sim \int |V_{kk'}|^2 \left[1 + \frac{\xi\xi' + \Delta^2}{\varepsilon\varepsilon'}\right] (f - f')\,\delta(\varepsilon' - \varepsilon - \omega) + \\ + \left(1 - \frac{\xi\xi' + \Delta^2}{\varepsilon\varepsilon'}\right)(1 - f - f')\,\delta(\varepsilon' + \varepsilon - \omega)]\, dk \approx \\ \approx \int |V_{kk'}|^2 \left(1 + \frac{\xi\xi' + \Delta^2}{\varepsilon\varepsilon'}\right)(f - f')\,\delta(\varepsilon' - \varepsilon - \omega)\, dk, \\ \xi' \equiv \xi(k'), \quad \varepsilon' \equiv \varepsilon(\xi'), \quad f = (e^{\varepsilon/T} + 1)^{-1}, \quad f' \equiv f(\varepsilon'), \end{aligned} \tag{3.64}$$

(ω is a quantity of the order of hyperfine splitting, $\omega \ll \Delta$; we therefore neglect pair creation in (3.64)). The ratio of the relaxation rates in the superconducting and the normal states is /1, 5, 249/

$$\frac{\alpha^s}{\alpha^n} \approx \frac{1}{\omega} \int \frac{[\varepsilon(\varepsilon + \omega) + \Delta^2]\,[f(\varepsilon) - f(\varepsilon + \omega)]\, d\varepsilon}{[(\varepsilon^2 - \Delta^2)\,((\varepsilon + \omega)^2 - \Delta^2)]^{1/2}}. \tag{3.65}$$

Measurements made by Hebel and Slichter /249/ in a zero field in Al are in good agreement with the theories if we assume that the quasi-particle energy levels have a spacing of the order of 0.01 T_c (Figure 28). (In actual fact, the theoretical curve is somewhat lower than in Figure 28 /250/.) The latest results for Al of Redfield and Anderson /251/ (Figure 29), Masuda and Redfield /252/, and Redfield /253/; of Masuda /254/ for Cd; of Hammond and Knight /255/, Hammond and Kelly /256/ for Ga; of Reif /258/ for Hg; and of Androes and Knight /259/ for Sn, also agree with theory. The corrected curves in Figure 29 correspond to $2\Delta = 3.25\ T_c$.

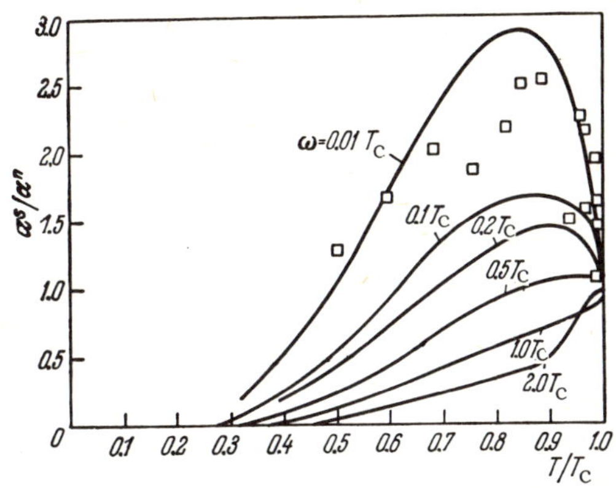

FIGURE 28. Ratio of rates of nuclear spin relaxation in superconducting and normal states. The experimental points /249, 251/ are the reciprocals of spin relaxation times in aluminum. The curve for $\omega = 0.01\ T_c$ was calculated by Hebel /249/ for widely spaced levels. The other curves were obtained by Miller /265/ for transfer of energy equal to ω.

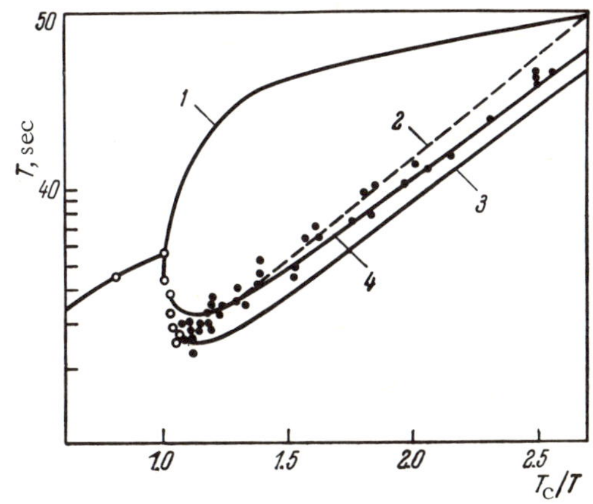

FIGURE 29. Nuclear relaxation time T_1 for superconducting aluminum. The theoretical curves were constructed on the basis of equation (3.65), with the density of states near the edges of the energy gap replaced by a function in the form of a rectangle of width $2d$ and height $d/2$, $d = \Delta/r$. The broken and solid curves were calculated for $2\Delta(0)/T_c$, equal to 3.52 and 3.25, respectively (the latter value was obtained in /283/). The experimental points correspond to the following measurements:

open circles) Redfield and Anderson /251/; dark circles) Redfield /253/, Masuda and Redfield /252/. Theoretical curves: 1) normal state ($H = H_c$) ; 2) to $r = 15$ (not corrected); 3) $r = 60$ (corrected); 4) $r = 15$ (corrected).

The width of the levels, introduced by Hebel and Slichter, is usually associated with anisotropy /252, 253/. This assumption seems to be confirmed by the experiments cited above and also by Masuda /257/, who showed that the introduction of impurities in Al decreases the width of the levels as we know, anisotropy decreases upon doping of superconductors) (see /258, 259/ and Chapter III, Sec. 2). However, Fibich /250/ showed that the interaction of electrons with thermal phonons may also lead to a similar broadening of levels.

The nuclear spin relaxation time T_1 for vanadium in the mixed state was measured by Pincus et al. and Fite and Redfield /260/, for Nb in the mixed state by Delrieu and Winter /261/, Asayama and Masuda /262/ (dependence of T_1 on the magnetic field). In /317/ nuclear spin relaxation was theoretically investigated in superconductors of dimensions small in comparison to the coherence length ξ_0. A noticeable contrast was noted in comparison with the case of bulk superconductors when spin-orbital interaction of electrons with lattice heterogeneities is taken into account. Reference /263/ contains a theoretical review of nuclear resonance in type-II superconductors.

Chapter IV

SUPERCONDUCTORS IN A VARIABLE ELECTROMAGNETIC FIELD

1. BEHAVIOR OF SUPERCONDUCTORS IN A WEAK VARIABLE FIELD

1.1. Theory

The behavior of superconductors in a variable field was first studied theoretically by Mattis and Bardeen /5, 264, 265/ and Abrikosov, Gor'kov, and Khalatnikov /266/. In the sequel we shall follow the exposition of /4, 172, 260/.

We consider the expression for the current density operator in the presence of a variable field with vector potential $\boldsymbol{A}$ (the scalar potential φ is assumed equal to zero):

$$\tilde{\boldsymbol{j}} = \tilde{\boldsymbol{j}}_1 - \frac{e^2}{m} \boldsymbol{A}(x) \tilde{\psi}^+(x) \tilde{\psi}(x), \tag{4.1}$$

$$\tilde{\boldsymbol{j}}_1 = \frac{ie}{2m} (\nabla_{r'} - \nabla_r)_{r' \to r} \tilde{\psi}^+(x') \tilde{\psi}(x),$$

$$\hbar = c = 1, \; x = \{\boldsymbol{r}, t\}.$$

Here the operators $\tilde{\psi}, \tilde{\psi}^+$ are written in the Heisenberg representation and depend on the field.

To go over from the interaction representation to the Heisenberg representation one uses the equation

$$\tilde{\boldsymbol{j}} = S^{-1}(t) \boldsymbol{j}(t) S(t), \tag{4.2}$$

$$S = T \exp\left\{i \int_{-\infty}^{t} (\boldsymbol{j}(x), \boldsymbol{A}(x)) \, d^4x\right\}.$$

In the linear field approximation we find from (4.1) and (4.2)

$$\tilde{j}_\alpha(x) = \tilde{j}_{1\alpha}(x) - \frac{e^2}{m} A_\alpha \psi^+(x) \psi(x) + \\ + i \int_{-\infty}^{t} [j_{1\alpha}(x), j_{1\beta}(y)] A_\beta(y) \, d^4y.$$

The value of the current at a given temperature evidently has the form

$$\boldsymbol{j}(x) = \langle \tilde{\boldsymbol{j}}(x) \rangle = \sum_m e^{\frac{\Omega+\mu N_m - E_m}{T}} \langle m | \tilde{\boldsymbol{j}}(x) | m \rangle. \tag{4.3}$$

Introducing the notation

$$P^R_{\alpha\beta}(x-y) = \begin{cases} i\langle [j_{1\alpha}(x),\ j_{1\beta}(y)] \rangle & \text{if} \quad t_x > t_y, \\ 0 & \text{if} \quad t_x < t_y, \end{cases}$$

and remembering that $\langle \boldsymbol{j}_1 \rangle = 0$, we find from (4.3)

$$j_\alpha(x) = -\frac{e^2 N}{m} A_\alpha(x) + \int P^R_{\alpha\beta}(x-y) A_\beta(y)\, d^4y. \tag{4.4}$$

After transformation to Fourier components, we obtain from (4.4)

$$\begin{gathered} j_\alpha(\boldsymbol{k}, \omega) = -Q_{\alpha\beta}(\boldsymbol{k}, \omega) A_\beta(\boldsymbol{k}, \omega), \\ Q_{\alpha\beta}(\boldsymbol{k}, \omega) = \frac{e^2 N}{m} \delta_{\alpha\beta} - P^R_{\alpha\beta}(\boldsymbol{k}, \omega). \end{gathered} \tag{4.5}$$

If we consider the same problem using a temperature technique, replacing t by an imaginary time parameter τ, we obtain instead of (4.4)

$$\begin{gathered} j_\alpha(\boldsymbol{r}, \tau) = -\frac{e^2 N}{m} A_\alpha(\boldsymbol{r}, \tau) + \int d\boldsymbol{r}_y \int_0^\beta d\tau_y \mathscr{P}_{\alpha\beta}(x-y), \\ \mathscr{P}_{\alpha\beta}(\boldsymbol{r}-\boldsymbol{r}', \tau-\tau') = \langle T\, [j_{1\alpha}(\boldsymbol{r}, \tau),\ j_{1\beta}(\boldsymbol{r}', \tau')] \rangle. \end{gathered} \tag{4.6}$$

After transformation to Fourier components, we obtain (4.5) with $\overline{\omega}_n$ instead of ω and

$$Q_{\alpha\beta}(\boldsymbol{k}, \overline{\omega}_n) = \frac{e^2 N}{m} \delta_{\alpha\beta} - \mathscr{P}_{\alpha\beta}(\boldsymbol{k}, \overline{\omega}_n), \quad \overline{\omega}_n = 2n\pi T.$$

It is easy to show in the usual way that $\mathscr{P}_{\alpha\beta}(\boldsymbol{k}, \overline{\omega}_n)$ is the analytic continuation of the function $P^R_{\alpha\beta}$, which is analytic in the upper half-plane, from the real axis to points on the imaginary axis, with ω replaced by $i\overline{\omega}_n$. When going from $P^R_{\alpha\beta}$ to $\mathscr{P}_{\alpha\beta}$ one must replace $\overline{\omega}_n$ by $-i(\omega + i\delta)$. Consequently, we can determine $\mathscr{P}_{\alpha\beta}(\boldsymbol{k}, \overline{\omega}_n)$ and $\boldsymbol{j}(\boldsymbol{k}, \overline{\omega}_n)$ by the temperature technique and then find their analytic continuations, and $P^R_{\alpha\beta}$ should have no singularities in the upper half-plane.

In order to find an expression for $\boldsymbol{j}(\boldsymbol{k}, \overline{\omega}_n)$, we formally consider the equation for the usual and the pairing temperature Green's functions G and $\mathcal{F}$ in a τ-dependent field $\boldsymbol{A}(\boldsymbol{r}, \tau) = \boldsymbol{A}(\boldsymbol{k}, \overline{\omega}_n) e^{i\boldsymbol{kr} - i\omega_n\tau}$:

$$\begin{gathered}\left[-\frac{\partial}{\partial\tau} + \frac{1}{2m}(\nabla_r - ie\boldsymbol{A}(\boldsymbol{r}, \tau))^2 + \mu\right] G(\boldsymbol{r}\tau, \boldsymbol{r}'\tau') + \\ + \Delta(\boldsymbol{r}\tau)\mathcal{F}^+(\boldsymbol{r}\tau, \boldsymbol{r}'\tau') = \delta(\boldsymbol{r} - \boldsymbol{r}'), \\ \left[\frac{\partial}{\partial\tau} + \frac{1}{2m}(\nabla_r + ie\boldsymbol{A}(\boldsymbol{r}, \tau))^2 + \mu\right]\mathcal{F}^+(\boldsymbol{r}\tau, \boldsymbol{r}'\tau') - \\ - \Delta^*(\boldsymbol{r}\tau) G(\boldsymbol{r}\tau, \boldsymbol{r}'\tau') = 0.\end{gathered} \tag{4.7}$$

We may assume that $G = G_0 + G'$; $\mathcal{F} = \mathcal{F}_0 + \mathcal{F}'$, where G_0 and $\mathcal{F}_0$ are the Green's functions in the absence of the field, and G' and $\mathcal{F}'$ are increments linear in the field. Then, from (4.7), we find

$$\begin{gathered}\left\{-\frac{\partial}{\partial\tau} + \frac{\nabla^2}{2m} + \mu\right\} G'(\boldsymbol{r}\tau, \boldsymbol{r}'\tau') + \Delta_0\mathcal{F}^{+\prime}(\boldsymbol{r}\tau, \boldsymbol{r}'\tau) = \\ = -\Delta'(\boldsymbol{r})\mathcal{F}_0^+(\boldsymbol{r} - \boldsymbol{r}', \tau - \tau') + \\ + \frac{ie}{2m}(\nabla, \boldsymbol{A} + \boldsymbol{A}, \nabla) G_0(\boldsymbol{r} - \boldsymbol{r}'. \tau - \tau'), \\ \left\{\frac{\partial}{\partial\tau} + \frac{\nabla^2}{2m} + \mu\right\}\mathcal{F}^{+\prime}(\boldsymbol{r}\tau, \boldsymbol{r}'\tau') - \Delta_0 G'(\boldsymbol{r}\tau, \boldsymbol{r}'\tau') = \\ = \Delta^{*\prime}(\boldsymbol{r}) G_0(\boldsymbol{r} - \boldsymbol{r}', \tau - \tau') - \\ - \frac{ie}{2m}(\nabla, \boldsymbol{A} + \boldsymbol{A}, \nabla)\mathcal{F}_0^+(\boldsymbol{r} - \boldsymbol{r}', \tau - \tau'), \quad \mathcal{F}_0^+ = \mathcal{F}_0.\end{gathered} \tag{4.8}$$

Since systems (4.7) and (4.8) are gauge-invariant, we may assume that $\boldsymbol{j}$ depends only on the transverse portion of the vector potential. We therefore assume that $\operatorname{div}\boldsymbol{A} = 0$. Δ' is an unknown function of $\boldsymbol{A}$; however, in the linear field approximation Δ', as a scalar, may depend only on $\operatorname{div}\boldsymbol{A}$, and if $\operatorname{div}\boldsymbol{A} = 0$ then also $\Delta' = 0$.

The solution of system (4.8) for G' has the form

$$\begin{gathered}G'(x, x') = \frac{ie}{m}\int [G_0(x - y)(\boldsymbol{A}(y), \nabla) G_0(y - x') + \\ + \mathcal{F}_0(y - x)(\boldsymbol{A}(y), \nabla)\mathcal{F}_0^+(x' - y)]\, d\boldsymbol{r}_y\, d\tau_y, \qquad x = \{\boldsymbol{r}, \tau\}.\end{gathered}$$

We transform to Fourier components and substitute the expression for G' into (4.6), using (4.1):

$$\begin{gathered}\boldsymbol{j}(\boldsymbol{k}, \overline{\omega}) = -\frac{2e^2T}{(2\pi)^3 m^2}\sum_{\omega'}\int \boldsymbol{p}(\boldsymbol{p}, \boldsymbol{A}(\boldsymbol{k}, \overline{\omega}))[G_0(p_+) G_0(p_-) + \\ + \mathcal{F}_0(p_+)\mathcal{F}_0^+(p_-)]\, d\boldsymbol{p} - \frac{Ne^2}{m}\boldsymbol{A}(\boldsymbol{k}, \overline{\omega})\Big] = -\frac{Ne^2}{m}\overline{Q}\boldsymbol{A}(\boldsymbol{k}, \overline{\omega}), \\ \omega \equiv \omega_n = (2n + 1)\pi T, \quad \omega' = \omega_{n'}, \quad p_\pm = \{\boldsymbol{p} \pm \boldsymbol{k}/2, \ \omega' \pm \overline{\omega}/2\}, \\ Q(\boldsymbol{k}, \overline{\omega}) = (Ne^2/m)\overline{Q}(\boldsymbol{k}, \overline{\omega}).\end{gathered}$$

$\boldsymbol{A}(\boldsymbol{r}, \tau)$ and $\boldsymbol{j}(\boldsymbol{r}, \tau)$ penetrate the superconductor only up to $\delta \sim 10^{-5}\,\mathrm{cm} \gg 1/p_F$, therefore the only significant terms in (4.9) are $\boldsymbol{A}(\boldsymbol{k}, \bar{\omega})$ and $\boldsymbol{j}(\boldsymbol{k}, \bar{\omega})$ for $k \sim 1/\delta \ll p_F$. Hence only values of $|\boldsymbol{p}| - p_F$ of the order of $|\boldsymbol{k}| \sim 1/\delta$ make a significant contribution to the integral in (4.9). Only two vectors $\boldsymbol{k}$ and $\boldsymbol{A}(\boldsymbol{k}, \bar{\omega})$ appear in (4.9), and $(\boldsymbol{k}, \boldsymbol{A}(\boldsymbol{k}, \bar{\omega})) = 0$. If we direct the z-axis in $\boldsymbol{p}$-space along $\boldsymbol{k}$ and integrate over the angle, it is readily seen that the vector $\boldsymbol{j}(\boldsymbol{k}, \bar{\omega})$ is directed along the vector $\boldsymbol{A}(\boldsymbol{k}, \bar{\omega})$. Substituting the expressions for G_0 and $\mathcal{F}_0$:

$$G_0 = -\frac{i\omega + \xi}{\omega^2 + \xi^2 + \Delta^2}, \qquad \mathcal{F}_0 = \frac{\Delta}{\omega^2 + \xi^2 + \Delta^2},$$

and then adding and subtracting the corresponding expressions for a normal metal:

$$\bar{Q}_n(\boldsymbol{k}, \bar{\omega}) - 1 = \frac{3T}{4} \sum_{\omega'} \int d\xi \int_0^{\pi} \sin^3\theta \, d\theta \frac{1}{(i\omega_+ - \xi_+)(i\omega_- - \xi_-)} = -1$$

(when $|\boldsymbol{k}| \to 0$, $\xi_\pm \approx \xi \pm (\boldsymbol{v}, \boldsymbol{k})/2$), we find from (4.9), after integrating over ξ:

$$\bar{Q}(\boldsymbol{k}, \bar{\omega}) = \frac{3\pi T}{4} \sum_{\omega'} \int_{-1}^{1} (1 - \mu^2)\, d\mu \times$$
$$\times \left\{ \frac{i\left(\omega_+ + \sqrt{\omega_+^2 + \Delta^2}\right)\left[i\left(\omega_- + \sqrt{\omega_+^2 + \Delta^2}\right) - v|\boldsymbol{k}|\mu\right] + \Delta^2}{\sqrt{\omega_+^2 + \Delta^2}\left[\omega_-^2 + \Delta^2 + \left(v|\boldsymbol{k}|\mu - i\sqrt{\omega_+^2 + \Delta^2}\right)^2\right]} + \right.$$
$$\left. + \frac{i\left(\omega_- + \sqrt{\omega_-^2 + \Delta^2}\right)\left[i\left(\omega_+ + \sqrt{\omega_-^2 + \Delta^2}\right) + v|\boldsymbol{k}|\mu\right] + \Delta^2}{\sqrt{\omega_-^2 + \Delta^2}\left[\omega_+^2 + \Delta^2 + \left(v|\boldsymbol{k}|\mu + i\sqrt{\omega_-^2 + \Delta^2}\right)^2\right]} \right\}, \tag{4.10}$$
$$v \equiv v_F = p_F/m$$

Obviously, for the conductivity we have $\sigma = -iQ(\boldsymbol{k}, \omega)/\omega$, since $\boldsymbol{E} = i\omega\boldsymbol{A}$, $\omega = i\bar{\omega}_n$.

The most interesting case is a Pippard superconductor, with $v|\boldsymbol{k}| \gg (T_c, \bar{\omega}_n)$, i.e., $\delta \ll \xi_0 \sim v/\Delta$. Then the integrand has a sharp maximum at $\mu \approx T_c/v|\boldsymbol{k}| \ll 1$. Therefore μ is negligible in the numerator under the integral sign. The remaining expression in curly brackets decreases like $1/\mu$ when $\mu \gg (T_c/v|\boldsymbol{k}|, \omega'/v|\boldsymbol{k}|)$. Let us write the terms which decrease like $1/\mu$ explicitly:

$$\bar{Q}(\boldsymbol{k}, \bar{\omega}) = \frac{3\pi T}{4} \sum_{\omega'} \int_{-1}^{1} d\mu \times$$
$$\times \left\{ \frac{\Delta^2 - \left(\omega_+ + \sqrt{\omega_+^2 + \Delta^2}\right)\left(\omega_- + \sqrt{\omega_-^2 + \Delta^2}\right)}{\sqrt{\omega_+^2 + \Delta^2}\left[\omega_-^2 + \Delta^2 + \left(v|\boldsymbol{k}|\mu - i\sqrt{\omega_+^2 + \Delta^2}\right)^2\right]} + \right.$$
$$+ \frac{\Delta^2 - \left(\omega_- + \sqrt{\omega_-^2 + \Delta^2}\right)\left(\omega_+ + \sqrt{\omega_+^2 + \Delta^2}\right)}{\sqrt{\omega_-^2 + \Delta^2}\left[\omega_+^2 + \Delta^2 + \left(v|\boldsymbol{k}|\mu + i\sqrt{\omega_-^2 + \Delta^2}\right)^2\right]} -$$

$$-i\frac{\omega_+ + \sqrt{\omega_+^2+\Delta^2}}{\sqrt{\omega_+^2+\Delta^2}\left[v\,|\boldsymbol{k}|\,\mu - i\left(\sqrt{\omega_+^2+\Delta^2}+\sqrt{\omega_-^2+\Delta^2}\right)\right]}+$$
$$+\,i\frac{\omega_- + \sqrt{\omega_-^{2}+\Delta}}{\sqrt{\omega_-^2+\Delta^2}\left[v\,|\boldsymbol{k}|\,\mu + i\left(\sqrt{\omega_+^2+\Delta^2}+\sqrt{\omega_-^2+\Delta^2}\right)\right]}\Bigg\}, \tag{4.11}$$

and integrate over μ. Then, letting $v\,|\boldsymbol{k}|\to\infty$, we find

$$\bar{Q}(\boldsymbol{k},\bar{\omega}) = \frac{3\pi^2 T}{4v\,|\boldsymbol{k}|}\sum_{\omega'}\left[1+\frac{\Delta^2-\omega'(\omega'-\bar{\omega})}{\sqrt{(\omega')^2+\Delta^2}\,\sqrt{(\omega'-\bar{\omega})^2+\Delta^2}}\right]. \tag{4.12}$$

The sum over ω' can be expressed as an integral:

$$\bar{Q}(\boldsymbol{k},\bar{\omega}) = \frac{3\pi i}{16v\,|\boldsymbol{k}|}\int\limits_C \mathrm{tg}\,\frac{\omega'}{2T}\left\{1+\right.$$
$$\left.+\frac{\Delta^2-\omega'(\omega'-\bar{\omega})}{\sqrt{(\omega')^2+\Delta^2}\,\sqrt{(\omega'-\bar{\omega})^2+\Delta^2}}\right\}d\omega', \tag{4.13}$$

where the contour C consists of two parts C_+ and C_-, as shown in Figure 30. The functions $\sqrt{(\omega')^2+\Delta^2}$ and $\sqrt{(\omega'-\bar{\omega})^2+\Delta^2}$ assume pure imaginary values on the slits. On the right edge of the upper slit and the left edge of the lower slit the imaginary parts of these functions are positive.

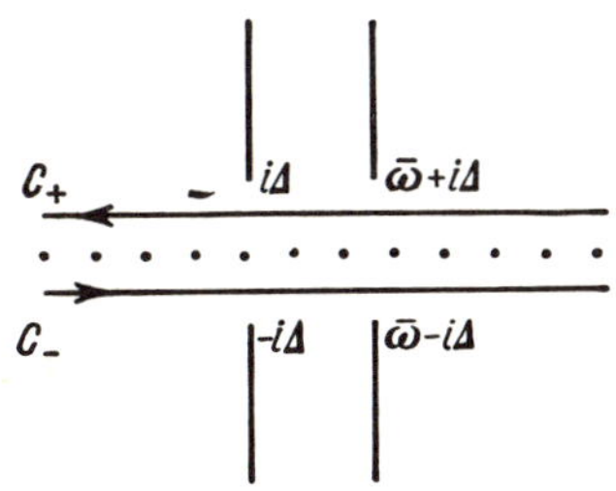

FIGURE 30. Integration contour for the integral in equation (4.13)

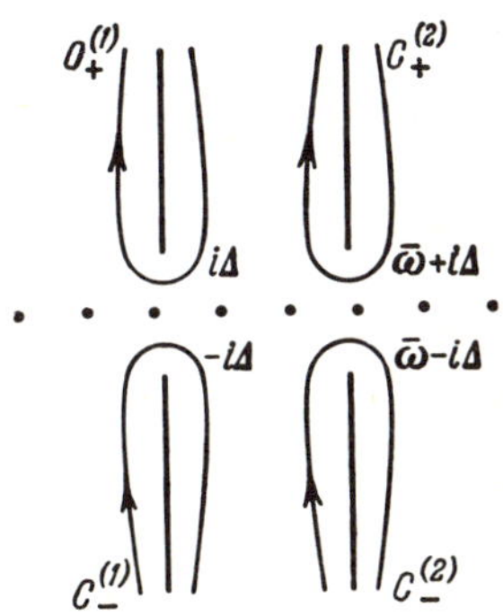

FIGURE 31. Transformation of the integration contour of Figure 30

We now go over from integration over the contours C_+ and C_- to the integration over $C_+^{(1)}$, $C_+^{(2)}$, $C_-^{(1)}$, and $C_-^{(2)}$ (Figure 31). When $\bar{\omega}_n = (2n+1)\pi T$ the integrals over $C_+^{(2)}$ and $C_-^{(2)}$ possess singularities, since the integration contour then passes through the point $\omega' = (2n+1)\pi T$, at which $\mathrm{tg}(\omega'/2T)$ is infinite. Therefore, in order to determine a branch of the function which is analytic in the upper half-plane of the variable $\omega = i\overline{\omega}_n$, we transform the integral so that for arbitrary values of ω the integration contour does not pass through singular points of the integrand. To this end, we use the periodicity of $\mathrm{tg}\,(\omega'/2T)$: for $\overline{\omega}_n = 2n\pi T$, the integral over $C_+^{(1)}$ equals the integral over $C_-^{(2)}$, and the integral over $C_-^{(1)}$ equals the integral over $C_+^{(2)}$. This is easily seen by substituting, say, $\omega' - \overline{\omega} = -u$ in the integral over $C_-^{(2)}$. Therefore (4.13) can be written

$$\overline{Q}(\boldsymbol{k}, \overline{\omega}) = \frac{3\pi i}{8v\,|\boldsymbol{k}|}\left(\int\limits_{C_-^{(1)}} + \int\limits_{C_+^{(1)}}\right)\left\{\mathrm{tg}\,\frac{\omega'}{2T}\left[1 + \frac{\Delta^2 - \omega'(\omega' - \overline{\omega})}{\sqrt{(\omega')^2 + \Delta^2}\,\sqrt{(\omega' - \overline{\omega})^2 + \Delta^2}}\right]\right\}d\omega'. \tag{4.14}$$

This expression is analytic as a function of $\omega = i\overline{\omega}_n$ in the upper half-plane of ω, since the integration contour does not go through any singular points of the integrand for $\mathrm{Im}\,\omega > 0$.

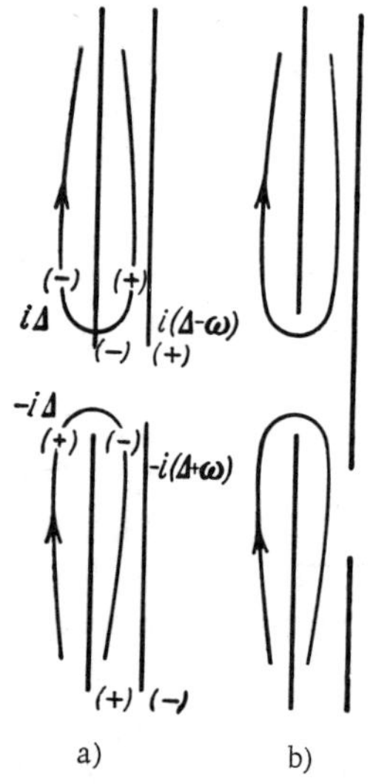

FIGURE 32. Integration contours for calculation of $\overline{Q}\,(\boldsymbol{k}, \omega)$ in different cases:

a) $\omega < 2\Delta$; b) $\omega > 2\Delta$.

An expression for $\overline{Q}(\boldsymbol{k}, \omega), \omega > 0$, is easily found, using Figure 32 for the cases $\omega < 2\Delta$ and $\omega > 2\Delta$ (in Figure 32 the choice of sign for the imaginary parts of the functions on the different edges of the slits is indicated in parentheses):

When $\omega < 2\Delta$:

$$\bar{Q}(\boldsymbol{k}, \omega) = \frac{3\pi}{4v|\boldsymbol{k}|}\Bigg\{\int\limits_{\Delta}^{\Delta+\omega} \operatorname{th}\frac{\omega'}{2T} \times$$

$$\times \frac{\omega'(\omega'-\omega)+\Delta^2}{\sqrt{(\omega')^2-\Delta^2}\sqrt{\Delta^2-(\omega'-\omega)^2}} d\omega' +$$

$$+ i\int\limits_{\Delta}^{\infty}\left[\operatorname{th}\frac{\omega'}{2T} - \operatorname{th}\frac{\omega'+\omega}{2T}\right] \times$$

$$\times \frac{\omega'(\omega'+\omega)+\Delta^2}{\sqrt{(\omega')^2-\Delta^2}\sqrt{(\omega'+\omega)^2-\Delta^2}} d\omega'\Bigg\}, \tag{4.15a}$$

and when $\omega > 2\Delta$:

$$\bar{Q}(\boldsymbol{k}, \omega) =$$

$$= \frac{3\pi}{4v|\boldsymbol{k}|}\Bigg\{\int\limits_{\omega-\Delta}^{\omega+\Delta} \operatorname{th}\frac{\omega'}{2T} \frac{\omega'(\omega'-\omega)+\Delta^2}{\sqrt{(\omega')^2-\Delta^2}\sqrt{\Delta^2-(\omega'-\omega)^2}} d\omega' +$$

$$+ i\int\limits_{\Delta}^{\omega-\Delta} \operatorname{th}\frac{\omega'}{2T} \frac{\omega'(\omega'-\omega)+\Delta^2}{\sqrt{(\omega')^2-\Delta^2}\sqrt{(\omega'-\omega)^2-\Delta^2}} d\omega' +$$

$$+ i\int\limits_{\Delta}^{\infty}\left[\operatorname{th}\frac{\omega'}{2T} - \operatorname{th}\frac{\omega'+\omega}{2T}\right] \times$$

$$\times \frac{\omega'(\omega'+\omega)+\Delta^2}{\sqrt{(\omega')^2-\Delta^2}\sqrt{(\omega'+\omega)^2-\Delta^2}} d\omega'\Bigg\}. \tag{4.15b}$$

The imaginary part of $\bar{Q}(\boldsymbol{k}, \omega)$ (i.e., the absorption) is different from zero only when $\omega > 2\Delta$.

We consider the empirical ratio of the impedance in the superconducting state $Z(\omega)$ to the real part of the impedance in the normal state R^n /4, 5/. This ratio is the same in the case of mirror and diffuse reflection. In the case of the half-space $z > 0$,

$$Z = \frac{E(0)}{\int\limits_0^\infty j\,dz} = -\frac{4\pi E(0)}{H(0)} = -\frac{4\pi i\omega A(0)}{H(0)}, \quad E(0) \equiv E_{z=0}.$$

For mirror reflection

$$Z = -\frac{4i\omega}{3\sqrt{3}}\sqrt[3]{\frac{-1}{Q(\boldsymbol{k}, \omega)\,k}};$$

for diffuse reflection

$$Z = -4\pi^2 i\omega \left\{ \int_0^\infty dk \ln\left[1 + 4\pi Q(k, \omega) k^{-2}\right] \right\}^{-1}.$$

In the Pippard case

$$\frac{Z}{R^n} \equiv \frac{R + iX}{R^n} = -2i \sqrt[3]{\frac{\pi\omega}{\Delta \widetilde{Q}(\omega)}} =$$

$$= 2\left[\frac{\pi\omega}{\Delta |\widetilde{Q}(\omega)|}\right]^{1/3} \left\{\sin\left(\frac{1}{3}\operatorname{arctg}\frac{\operatorname{Im}\widetilde{Q}}{\operatorname{Re}\widetilde{Q}}\right) - \right.$$

$$\left. - i\cos\left(\frac{1}{3}\operatorname{arctg}\frac{\operatorname{Im}\widetilde{Q}}{\operatorname{Re}\widetilde{Q}}\right)\right\}, \qquad (4.16)$$

$$\widetilde{Q} = \frac{4v|k|}{3\Delta} \bar{Q}(k, \omega).$$

When $T = 0$

$$\widetilde{Q}_0(\omega) = \begin{cases} 2\pi E\left(\frac{\widetilde{\omega}}{2}\right) & \text{if} \quad \widetilde{\omega} < 2, \\ \pi\left[\widetilde{\omega} E\left(\frac{2}{\widetilde{\omega}}\right) - \frac{\widetilde{\omega}^2 - 4}{\widetilde{\omega}} K\left(\frac{2}{\widetilde{\omega}}\right)\right] + i\pi\Theta(\widetilde{\omega} - 2) \times \\ \times\left[\widetilde{\omega} E\left(\sqrt{1 - 4/\widetilde{\omega}^2}\right) - \frac{4}{\widetilde{\omega}} K\left(\sqrt{1 - 4/\widetilde{\omega}^2}\right)\right] & \text{if} \quad \widetilde{\omega} > 2, \end{cases} \qquad (4.17)$$

where $\widetilde{\omega} = \omega/\Delta$,

$$\Theta(x) = \begin{cases} 1 & \text{if} \quad x > 0, \\ 0 & \text{if} \quad x < 0, \end{cases}$$

$E(x)$ and $K(x)$ are complete elliptic integrals.

The dependence of R/R^n and X/R^n on $\widetilde{\omega}$ at $T = 0$ is shown in Figure 33.

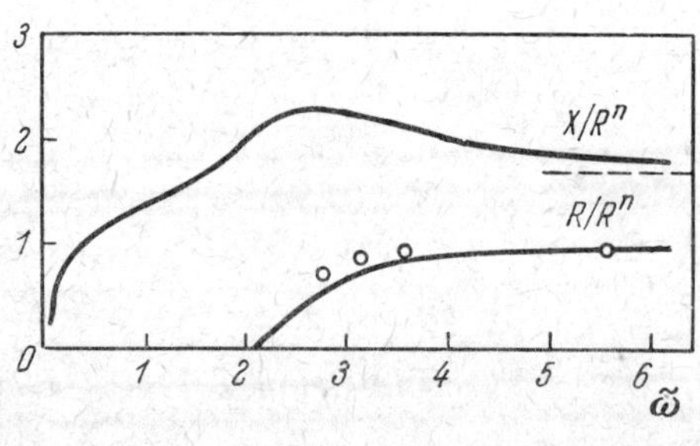

FIGURE 33. Theoretical dependence of R/R^n and X/R^n at $T = 0$ on frequency, according to equation (4.17)

We now find Z/R^n in various particular cases, $T \neq 0$.

a) $\omega \ll \Delta(0) \equiv \Delta_{T=0}$. If the inequality $T \ll \omega$ is also satisfied, equation (4.17) and Figure 33 can be used.

In the region $\omega \ll T \ll \Delta$ we find from (4.15) and (4.16)

$$\frac{Z(\omega)}{R^n} = 2\left(\frac{\omega}{\pi\Delta}\right)^{1/3}\left\{\frac{4}{3}\operatorname{sh}\frac{\omega}{2T}K_0\left(\frac{\omega}{2T}\right)e^{-\Delta/T} - i\left[1+\frac{1}{3}\left(\frac{\omega}{4\Delta}\right)^2+\frac{2}{3}e^{-\omega/2T}I_0\left(\frac{\omega}{2T}\right)e^{-\Delta/T}\right]\right\}. \tag{4.18}$$

In the region $\omega \ll \Delta \ll T$, $\Delta/T \gg \omega/\Delta$

$$\frac{Z(\omega)}{R^n} = 2\left[\frac{\omega}{\pi\Delta\operatorname{th}(\Delta/2T)}\right]^{1/3}\left\{\frac{2}{3\pi}\frac{\omega}{T\operatorname{sh}(\Delta/T)}\times \ln\left(2\sqrt{\frac{2\Delta}{\omega}}\right)+\frac{1}{3\pi}\frac{\omega}{\Delta}\left(\operatorname{cth}\frac{\Delta}{2T}-1\right) - \frac{2}{3\pi}\frac{\omega}{T}P\left(\frac{\Delta}{T}\right)\operatorname{cth}\frac{\Delta}{2T} - i\left[1+\frac{1}{3}\left(\frac{\omega}{4\Delta}\right)^2\right]\right\}, \tag{4.19}$$

where

$$P(x) = \int_1^\infty \frac{d\varepsilon}{(\varepsilon^2-1)}\,\frac{\operatorname{ch}x\varepsilon-\operatorname{ch}x}{(\operatorname{ch}x\varepsilon+1)(\operatorname{ch}x+1)} = \begin{cases} e^{-x}\ln 2\gamma x & \text{if} \quad x \gg 1, \\ \left(\frac{7}{2\pi^2}\right)\zeta(3)\,x & \text{if} \quad x \ll 1. \end{cases}$$

If the condition $\Delta/T \gg \omega/\Delta$ is not satisfied, we use the inequality $\Delta \ll T$ to obtain

$$\frac{Z(\omega)}{R^n} = -2i\left(-i+\frac{\pi\Delta^2}{2T\omega}\right)^{-1/3}. \tag{4.20}$$

b) $\omega \sim \Delta(0)$. In this case it is difficult to simplify the expression for Z/R^n without additional approximations.

If $T \ll \omega \sim \Delta$ and $\omega < 2\Delta(0)$, then $\widetilde{Q}(\omega)$ differs only slightly from $\operatorname{Re}\widetilde{Q}(\omega)$ at $T=0$, i. e., from $2\pi E(\omega/(2\Delta))$, and

$$\frac{Z(\omega)}{R^n} = 2\left[\frac{\omega}{2\Delta E(\omega/2\Delta)}\right]^{1/3}\left\{\frac{e^{-\Delta/T}}{3E(\omega/2\Delta)}\left[\pi T\left(\frac{1}{\omega}+\frac{1}{2\Delta}\right)\right]^{1/2} - i\left[1+\frac{e^{-\Delta/T}}{3E(\omega/2\Delta)}(\pi T)^{1/2}\left(\frac{1}{\omega}-\frac{1}{2\Delta}\right)^{1/2}\right]\right\}. \tag{4.21}$$

If $T \ll \omega \sim \Delta$ and $\omega > 2\Delta(0)$, both parts of $\widetilde{Q}(\omega)$ are of the same order, and

$$\widetilde{Q}(\omega, T) = \widetilde{Q}_0(\omega) + 2\pi e^{-\frac{(\omega-\Delta)}{T}} \sqrt{\pi T\left(\frac{1}{2\Delta} - \frac{1}{\omega}\right)} - \\ - 2\pi i e^{-\Delta/T} \sqrt{\pi T}\left[\sqrt{\frac{1}{\omega} + \frac{1}{2\Delta}} - \sqrt{\frac{1}{2\Delta} - \frac{1}{\omega}}\right]. \quad (4.22)$$

We find Z/R^n from (4.16).

c) $\omega \gg \Delta(0)$. In this case the real part of $\widetilde{Q}(\omega)$ is small in comparison with the imaginary part and

$$\frac{Z(\omega)}{R^n} = 1 + \left(\frac{\Delta}{\omega}\right)^2\left[\frac{2}{3}\ln\frac{2\omega}{\Delta(0)} + \frac{1}{3} - \frac{\pi}{\sqrt{3}}\right] - \\ - i\sqrt{3}\left[1 + \left(\frac{\Delta}{\omega}\right)^2\left(\frac{2}{3}\ln\frac{2\omega}{\Delta(0)} + \frac{1}{3} - \frac{\pi}{3\sqrt{3}}\right)\right]. \quad (4.23)$$

Simple estimates show that the London case (when the magnetic field penetration depth is significantly larger than the coherence length) may be observed in pure superconductors only very close to the transition temperature T_c and at frequencies markedly less than T_c; this case is therefore not very interesting.

A theoretical investigation of the behavior of superconductors with nonmagnetic impurities in a weak high-frequency field has been made by Abrikosov and Gor'kov /100/ (see also /267/). It was assumed in /100/ that the impurity concentration was not enough to modify the electron-electron interaction; even in this case, however, the free path l for scattering of electrons off impurities may be less than the size of a Cooper pair $\xi_0 = v/\Delta$. Then l will serve as the correlation parameter rather than ξ_0. (The exact calculation shows that the coherence length is $\sqrt{\xi_0 l}$ when $l < \xi_0$.) Therefore, when $\sqrt{\xi_0 l} < \delta$ we can expect to see local coupling between the current $\boldsymbol{j}$ and the vector potential $\boldsymbol{A}$, i.e., when doped a Pippard superconductor becomes a London superconductor (δ is the penetration depth).

Here we present only the final results of reference /100/. If $U(\boldsymbol{p} - \boldsymbol{p}')$ is the Fourier component of the interaction potential between an electron and an impurity, then the reciprocal of the so-called "transport time" of the free path, τ_{tr} is determined by the integral

$$\frac{1}{\tau_{tr}} = \frac{nmp_F}{(2\pi)^2}\int |U(\theta)|^2(1 - \cos\theta)\,d\Omega,$$

where $U(\theta) = U(\boldsymbol{p} - \boldsymbol{p}')$ when $|\boldsymbol{p}| \sim |\boldsymbol{p}'| \sim p_F$, p_F is the Fermi momentum and n the density of impurity atoms. We consider the case $T = 0$ and $\tau_{tr}\Delta \ll 1$ $(l \ll \xi_0)$.

In the case $\omega < 2\Delta$

$$Q(\omega) = 2\sigma\Delta E\left(\frac{\omega}{2\Delta}\right), \tag{4.24}$$

where $\sigma = (Ne^2/m)\,\tau_{tr}$ is the impurity conductivity of a metal in the normal state; $E(x)$ and $K(x)$ are complete elliptic integrals.

In the case $\omega > 2\Delta$

$$Q(\omega) = \sigma\omega\left\{E(\varkappa) - \frac{\omega^2 - 4\Delta^2}{\omega^2}K(\varkappa) - i\left[E(\varkappa_1) - \frac{4\Delta^2}{\omega^2}K(\varkappa_1)\right]\right\}, \tag{4.25}$$

$$\varkappa = 2\Delta/\omega, \qquad \varkappa_1 = \sqrt{1-\varkappa^2};$$

and in the case $\omega \gg \Delta$

$$Q(\omega) = -\sigma\omega\left\{i\left[1 - \frac{2\Delta^2}{\omega^2}\left(\ln\frac{2\omega}{\Delta} + \frac{1}{2}\right)\right] - \frac{\pi\Delta^2}{\omega^2}\right\}. \tag{4.26}$$

If small terms of the order of Δ^2/ω^2 are dropped from (4.26), the relationship between the current and the field is the same as in a normal metal: $j = \sigma E$ (E is the electric field intensity).

Near the threshold $|\omega - 2\Delta| \ll \Delta$, we obtain

$$Q(\omega) = \sigma\left[2\Delta + \left(\Delta - \frac{\omega}{2}\right)\left(\ln\frac{16\Delta}{2\Delta - \omega} - 1\right)\right], \quad \omega < 2\Delta,$$

$$Q(\omega) = \sigma\left[2\Delta - \left(\frac{\omega}{2} - \Delta\right)\left(\ln\frac{16\Delta}{\omega - 2\Delta} - 1 + i\pi\right)\right], \quad \omega > 2\Delta. \tag{4.27}$$

From (4.27) we see that an imaginary component appears in $Q(\omega)$, i.e., in the current j, when $\omega > 2\Delta$. The impedance of a bulk specimen with a plane surface is $Z = -i\omega\,(4\pi/Q(\omega))^{1/2}$. Figure 34 illustrates the real and imaginary parts of $Q(\omega)$ (in units of $2\sigma\Delta/c$) as functions of $\omega/2\Delta$.

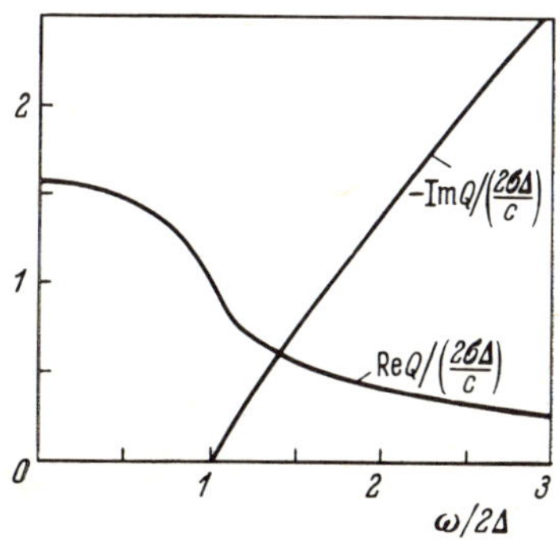

FIGURE 34. Theoretical dependence of Re $Q(\omega)/(2\sigma\Delta/c)$ and Im $Q(\omega)/(2\sigma\Delta/c)$ on $\omega/2\Delta$, according to equations (4.24) and (4.25)

Abrikosov and Gor'kov /199/ also investigated the behavior of an alloy with a magnetic impurity. In /199/ it was shown that, at a certain impurity concentration, namely $0.91\,n_c$, the gap in the energy spectrum disappears, although superconductivity is preserved (n_c is the concentration at which pairing disappears). The source of this gapless superconductivity is evidently that even at absolute zero some of the Cooper pairs are broken, owing to the unpairing effect of the magnetic field of the impurities. As expected, in the gapless region $Q(\omega)$ is the same as for a normal metal /199/:

$$Q_{\mathrm{gl}}(\omega) = \frac{1}{4\pi\delta^2} - i\sigma\omega. \tag{4.28}$$

The threshold absorption of electromagnetic waves was theoretically investigated by Pokrovskii and Ryvkin /268/.

An expression for $Q(\boldsymbol{k}, \omega)$ in the case of pure strong coupling superconductors was obtained by Medvedev /269/, who showed that formula (4.15) for $\bar{Q}(\boldsymbol{k}, \omega)$ must be supplemented by terms $\Delta\bar{Q}$ of the form $(T_c/\omega_D)^2$ (ω_D is the Debye frequency); this implies, naturally, that the energy gap Δ in the formula for $Q(\boldsymbol{k}, \omega)$ should be given a value conforming to the strong coupling theory rather than to the BCS theory /1, 5/. For example, for Pb at 0°K, $2\Delta/T_c = 4.1$ to 4.3, rather than 3.5. Near T_c,

$$\Delta(T) = \alpha T_c \sqrt{1 - T/T_c}$$

with $\alpha = 4$ rather than 3.06 /82–85/.

The dependence of the impedance on a stationary magnetic field in pure superconductors was calculated by Maki /270/. A complete theory of the behavior of films in the field of an electromagnetic wave with $d \ll \xi_0$ and $d^2/\xi_0 \ll l \ll \sqrt{d\xi_0}$, both without a stationary magnetic field and in its presence, was developed by Ovchinnikov /271/ (ξ_0 is the coherence length of the pure superconductor and d the film thickness). The impedance was calculated by Maki /272/ for highly doped films ($l \ll d$). Urushadze /273/ calculated the impedance for a film along which a current is flowing.

In deriving equation (4.15) for $\bar{Q}(\boldsymbol{k}, \omega)$, we did not allow for nonlinear effects (see Sec. 2).

The theory of surface impedance in type-II superconductors was developed by Caroli and Maki /274/, Maki /275/, Hibler and Cyrot /276/. In microwave absorption in type-II superconductors, excitation of vibrations of the Abrikosov vortex lattice by an electromagnetic wave is a major factor.

In /324/ the transmission through moderately thick films of electromagnetic radiation at a frequency ω significantly smaller than the energy gap Δ was calculated on the assumption that $l \gg v_F/\omega$, allowing for the emission of electron excitations from the superconducting skin (l is the free path for scattering off impurities and v_F the velocity of electrons on the Fermi surface).

In /325/ the impedance was calculated for heterogeneous superconductors, and the frequency and temperature dependence of the impedance specific for this case was found.

1. 2. Comparison with Experiment

The passage of infrared radiation through films of thickness smaller than the penetration depth was investigated experimentally by Glover and Tinkham /277/ and Ginsberg and Tinkham /278/. The ratio of the complex conductivities in the superconducting and normal states, $\sigma/\sigma^n = (\sigma_1 - i\sigma_2)/\sigma^n$, is a convenient index for comparison of theory with experiment.

The frequency dependence of σ_1/σ^n, σ_2/σ^n and the ratio of the transmissivities through a film in the superconducting and normal states, T and T^n, calculated by Tinkham /278/ and Miller /265/ via the theoretical equations of Mattis and Bardeen, are presented in Figure 35 ($\omega_g = 2\Delta$). The experiments confirm the presence of a peak in the T/T^n curve at $\omega \sim \omega_g$.

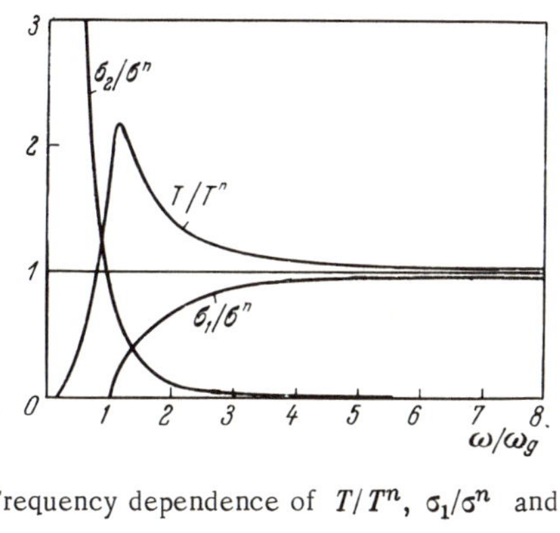

FIGURE 35. Frequency dependence of T/T^n, σ_1/σ^n and σ_2/σ^n /278/

Figure 36 presents $(1 - \sigma_1/\sigma^n)^{-1}$ on a logarithmic scale as a function of ω/T_c for tin films, according to /278/. The theory predicts an approximately straight line with slope ~ 1.65; the experimental slope is ~ 1.83.

Extrapolation to $\sigma_1 = 0$ yields a value for the energy gap width.

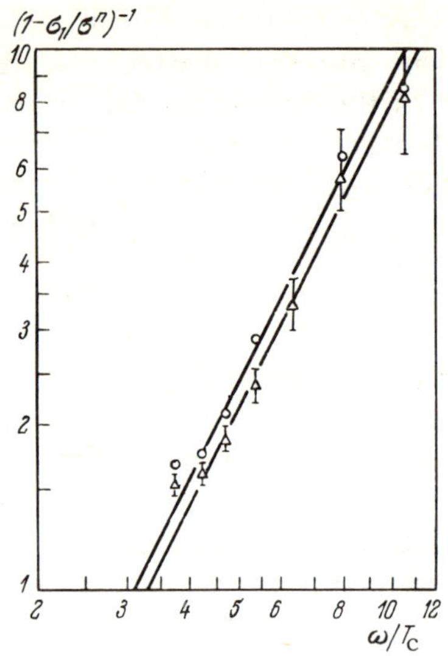

FIGURE 36. Frequency dependence of $(1 - \sigma_1/\sigma^n)^{-1}$ for tin /278/. Experimental points: triangles) Sn I (152 ohm/cm^2); circles) Sn II (230 ohm/cm^2).

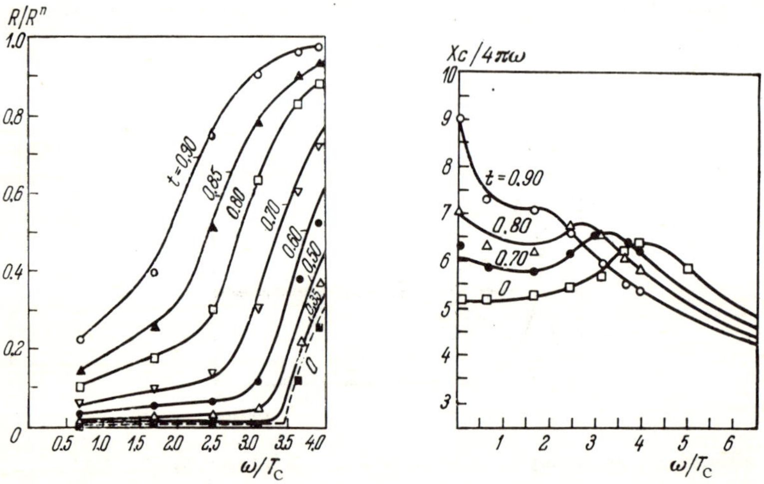

FIGURE 37. Frequency dependence of R/R^n and X in aluminum, from measurements in /283/

Thus, the experimental values of $2\Delta(0)/T_c$ found in /278/ were 4.0 ± 0.5, 3.3 ± 0.2, and 3.9 ± 0.3 for lead, tin, and indium respectively. In the first experiments /278, 279, 280/ for lead and mercury small preliminary absorption peaks were observed at a frequency somewhat lower than the primary peak. In later and more accurate experiments, however, the existence of the preliminary peaks was not confirmed either for lead films /281/ or for Pb bulk specimens /282/.

Careful measurements of surface impedance were made by Biondi and Garfunkel over a wide range of frequencies and temperatures /283/ (see also /284/). In Figure 37 R/R^n and X are shown as functions of ω/T_c for aluminum (solid lines) according to /283/ ($t = T/T_c$; X is expressed in terms of effective penetration depth $\delta_{eff} = cX/4\pi\omega$). The broken line and points correspond to Miller's numerical calculations using the theoretical equations for the Pippard case /264, 265/; here, however, the value 3.37, rather than 3.52, was taken for the quantity $2\Delta(0)/T_c$. One sees from Figure 37 that for R/R^n the experimental data deviate markedly from the theory of the limiting Pippard case. Figure 38 shows the dependence of R/R^n on T for Al, according to /284/.

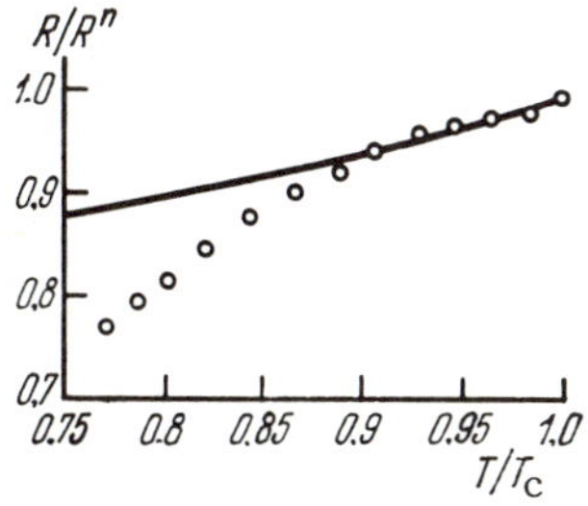

FIGURE 38. Temperature dependence of R/R^n for aluminum at $\omega = 3.04\ T_c$. Curve: theoretical function (4.23) /266/; points: experimental data of /284/.

Thompson /285/ measured the surface resistivity for lanthanum; the experimental values for R/R^n fit the theory. In recent measurements of the passage of infrared radiation through Pb films of thickness 50 to 150 Å, Lehoczky and Briscoe /286/ observed good agreement with the Mattis-Bardeen theory when $T < T_c$ but above T_c (near T_c) they found an increment to the impedance of a normal metal, associated with fluctuations in the superconducting component above T_c, in good agreement with the theoretical calculations of Schmidt and Mikeska /287/ (see also /322/).

Khaikin measured the surface impedance of single-crystal cadmium in the temperature range $0.4\,T_c$ to $0.9\,T_c$ at frequency $\omega = 0.9\ T_c$ /288/. From Figure 39 one sees that the experimental data fit the theoretical equation (4.15) quite well /266/. The experimental data do not fit the theory so well at low frequencies for mercury ($\omega = 0.014\ T_c$ /289/) and tin ($\omega = 0.155\ T_c$ /289/ and $\omega = 0.11\ T_c$ according to Prosorov's measurements; see /266/).

The theoretical value of R/R^n /266/ for mercury /289/ at $t = T/T_c = 0.6-0.7$ is about half that of the experimental value. This divergence increases near T_c, and at T_c, i.e., $T \approx 0.01°$, the theoretical value of R/R^n falls to one sixth of the experimental value, if we assume as in the BCS theory that near T_c we have $\Delta = \alpha T_c \sqrt{1 - T/T_c}$ with $\alpha = 3.06$. This discrepancy is even greater if we recall that for mercury $\alpha \approx 4.3$–4.6 /84, 85/.

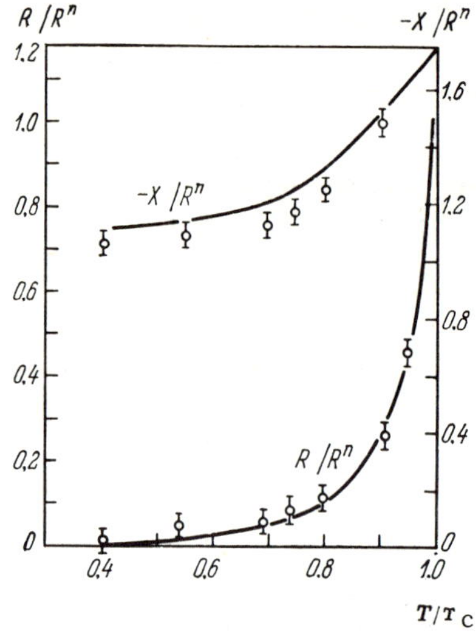

FIGURE 39. Temperature dependence of R/R^n and $-X/R^n$ for cadmium at $\omega = 0.9\ T_c$. Points: measurements of /288/; curve: theoretical equation (4.21) /266/.

The difference between theory and Pippard's experimental data for tin /289/ is also very significant. At somewhat higher frequencies ($\omega = 0.11\ T_c$) the discrepancy between Prozorov's measurements and theory is somewhat smaller, but is nevertheless high, expecially near T_c. At $T \approx 0.75\ T_c$ the experimental value of R/R^n is 30 percent higher than the calculated value and near T_c these quantities differ by a factor of three or four. The difference is somewhat smaller for the reactance in Prozorov's measurements: for $|X^n - X|/R^n$, it is at most two (see /266/). The experimental data for tin from Kaplan et al. /291/, Biondi, Forrester, and Garfunkel /292/ (R/R^n for polycrystalline tin) and Rudheimer et al. /293/ (reflection and absorption of microwave radiation by thin tin films) refer to a different frequency range. It should be noted that tin cannot be regarded unambiguously as a Pippard superconductor, belonging rather to an intermediate case; this may account partially for the difference between theory and experiment. However, for mercury the source of this discrepancy is not completely clear at the present time. Incidentally, it is noteworthy that in the frequency and temperature range considered Z should vary very rapidly and this greatly complicates the experiments. In fact, under the conditions

$$\omega \ll T_c \text{ and } \Delta^2 \ll \omega T$$

and very close to T_c, R/R^n increases, according to (4.20), like $(1 - T/T_c)^{-1/3}$, while even closer to T_c, Z begins to fall and approaches Z^n.

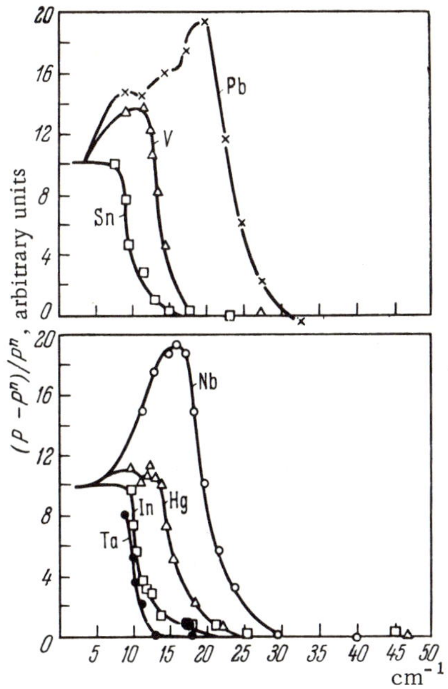

FIGURE 40. Frequency dependence of electromagnetic power absorbed by bulk superconducting specimens in the far infrared /279/:

values of T_c: Pb) 7.15; V) 5.10; Sn) 3.73; Nb) 9.00; Hg) 4.15; In) 3.99; Ta) 4.39.

Figure 40 gives results of measurements by Richards and Tinkham /279/ of the frequency dependence of electromagnetic power P absorbed by various superconductors in the far infrared; Figure 41 gives the absorption curve for V at different temperatures /279/. Values of Δ (0) for Al, Sn, In, Ta, V, Hg, Nb, Pb were found in these experiments. The spread of the points for Pb and Hg in Figure 40, which seems to indicate preliminary peaks, is attributed in /281, 282/ to experimental errors (see below). To investigate the effect of proximity, the surface impedance of a normal state tin film placed on a superconducting lead film was measured in /294/.

In experiments of Richards /295/ (frequency dependence of absorption in the far infrared), Adkins /296/ (surface resistivity in the microwave region), and Lewis /297/ (temperature dependence of R/R^n), anisotropy was detected in the energy gap of pure lead. Anisotropy was also found for pure Al and Zn by Biondi, Garfunkel, and Thompson /298/ (by the same method as in /205/), for pure zinc by Zemon and Boorse /299/, and for pure lanthanum by Leslie et al. /300/ (microwave radiation reflection). Norman /301/, investigating thick films of lead and tin by the infrared absorption method, found two peaks, due to either anisotropy or a multiband structure. Pokrovskii and Ryvkin /302/ showed that allowance for anisotropy should cause a 10 percent decrease in the energy gap values obtained from threshold absorption.

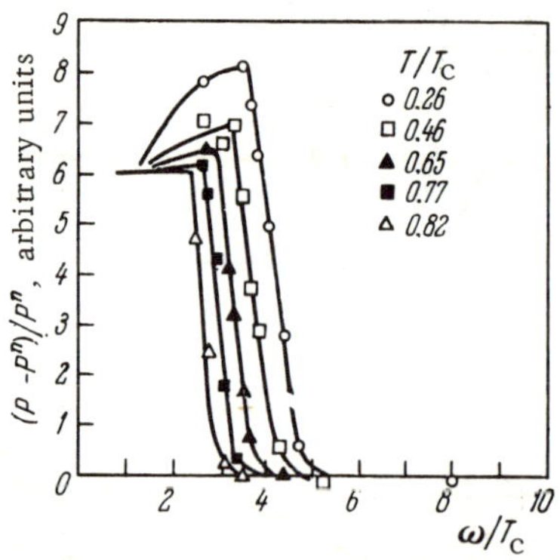

FIGURE 41. Dependence of absorbed electromagnetic power on frequency for vanadium at different temperatures /279/

In /295/ it was shown that the absorption edge becomes sharper, i.e., the energy gap anisotropy disappears, when tin is doped with indium. Similar results were obtained by Ginsberg and Leslie /303/ for Pb—Tl alloys with different Tl content. The critical temperature in Pb—Tl alloys was measured by Finnemore and Mapother /304/.

In the first experiments /278, 279/ the effect of a magnetic field on the impedance was not detected either in films or in bulk specimens. However, in a more careful investigation the effect was detected by Spiewak /305/, who found that $R(H) - R(0)$ and $X(H) - X(0)$ may have different signs, depending on the relative orientation of the stationary and variable fields and on the temperature. Budzinski and Garfunkel /306/ detected a drop in the absorption threshold for Al at $H = 0.8H_c$, which is clearly due to the decrease in the energy gap in the magnetic field; anisotropy of the energy gap in the presence of a field was also detected. Both these effects are greatly diminished by doping the specimen. White and Tinkham /807/ studied microwave absorption in films of In, Sn, and Pb in the presence of a magnetic field. Ovchinnikov's theoretical calculations /271/ explain these experiments quite well.

Experiments on surface impedance in type-II superconductors were performed by Hibler and Maxfield /308/. They observed a faster drop in the ratio R/R^n at H_{c2} than that predicted by the theory of Hibler and Cyrot /276/.

Impedance in rigid superconductors in the presence of pinning was theoretically and experimentally investigated in Gittleman and Rosenblum /309/ in a wide frequency range for Pb—In and Nb—Ta alloys.

2. STRONG FIELDS

The behavior of superconductors in weak fields (both constant and variable) was investigated soon after the appearance of the microscopic theory of superconductivity. We have already discussed the transient properties of superconductors in weak fields. In regard to strong variable fields, however, the situation is much more complicated.

The celebrated Ginzburg-Landau theory was substantiated with the aid of the microscopic theory in Gor'kov /327/. This theory describes the behavior of superconductors in strong stationary fields at near-critical temperatures. In regard to variable fields, many authors have suggested that the nonsteady-state electrodynamics is described by a simple generalization of the Ginzburg-Landau equations, by merely adding a term involving the time derivative of the ordering parameter. It was thus assumed that the transition to the description of nonsteady-state phenomena is similar to the transition from the stationary to the time-dependent Schrödinger equation. However, a detailed analysis of the question published recently by Gor'kov and Eliashberg /328—331/ showed that the pattern is not nearly so simple. It turns out that, generally speaking, there is no simple and universal generalization of the Ginzburg-Landau theory to the dynamic case.

As mentioned, Gor'kov /327/ presented a microscopic derivation of the Ginzburg-Landau equations, which describe the behavior of superconductors as $T \to T_c$ in an arbitrary stationary field. Gor'kov's equations yield an integral equation for $\Delta(\boldsymbol{r})$ by a suitable expansion in powers of Δ/T near T_c. The feasibility of such an expansion and the subsequent reduction of the integral equation to a differential equation is by no means obvious and, furthermore, is possible only in certain cases. In particular, it has turned out to be possible in an important case, corresponding to the region in which the Ginzburg-Landau theory is applicable. In the nonsteady-state case the kernels of the corresponding integral equations are not slowly decreasing functions, in contradistinction to the steady-state case. All the quantities possess singularities as functions of frequency, owing to threshold absorption at $\omega = 2\Delta$. Therefore, the equations for the time-dependent ordering parameter $\Delta(\boldsymbol{r}, t)$ are generally not local. The assumption that the dynamic superconductor state may be described by a complex amplitude $\Delta(t, \boldsymbol{r})$ (as in the equilibrium state) is not intuitively obvious.

An exceptional case is presented by superconducting alloys with paramagnetic impurities, which, as is known /199/, are characterized by gapless superconductivity at a sufficiently high concentration of impurities. For such alloys it has been found possible to simplify the general scheme of the theory considerably. Moreover, consideration of this case also yields several general conclusions.

Note, moreover, that in /328—331/, where the properties of superconductors in strong variable electric fields are investigated, the temperature range near the critical point was considered. It is in this range that one can isolate the Joule loss and develop a scheme independent of the experimental setup.

When the discussion is limited to finite temperatures, the equations of microscopic superconductivity theory /3, 172/ form a system of equations for the usual and anomalous Green's functions. The temperature Green's functions are then considered, i.e., the "time" variable is "imaginary" time. This approach enables us successfully to treat the equilibrium thermodynamic properties of superconductors. To describe time-dependent processes, however, one must perform analytic continuation onto the usual time axis.

In the general case, the space and time heterogeneities of the Green's function, $G_{\varepsilon_1\varepsilon_2}(\boldsymbol{k}_1, \boldsymbol{k}_2)$ and $\mathcal{F}^+_{\varepsilon_1\varepsilon_2}(\boldsymbol{k}_1, \boldsymbol{k}_2)$, depend on two imaginary frequencies and two momenta. The Fourier component of the superconducting ordering parameter $\Delta_\omega(\boldsymbol{k})$ is given by

$$\Delta^*_\omega(\boldsymbol{k}) = gT \sum_\varepsilon \int \frac{d\boldsymbol{p}}{(2\pi)^3} \mathcal{F}^+_{\varepsilon,\, \varepsilon-\omega}(\boldsymbol{p},\ \boldsymbol{p}-\boldsymbol{k}), \quad \varepsilon = i(2n+1)\pi T. \tag{4.29}$$

The quantity $\Delta(\boldsymbol{r}, t)$ is viewed as a self-consistent field acting on the electron system. Each term of the sum in (4.29) may be regarded as the residue of an integral along a contour around the point $z = \varepsilon$, so that

$$T \sum_\varepsilon \to \frac{1}{4\pi i} \int_C \operatorname{th} \frac{z}{2T}\, dz. \tag{4.30}$$

With the aid of the Gor'kov equations, the functions $G_{\varepsilon,\varepsilon-\omega}$ and $\mathcal{F}^+_{\varepsilon,\varepsilon-\omega}$ were expressed in /328/ as series in powers of Δ_ω and $\Delta_{\omega'}$. Each term of this series contains the factors $G^\pm(\varepsilon) = 1/(\varepsilon \pm \xi)$.

Next, transforming the sum in (4.29) into an integral in the complex time plane by means of (4.30) and transforming the integration contours into straight lines parallel to the imaginary axis we obtain expressions for the terms of the series, containing products of the advanced and retarded Green's functions, G^A and G^R, respectively. The former do not possess singularities in the lower half-plane, and the latter do not possess singularities in the upper half-plane. This enables us to perform analytic continuation from discrete imaginary points $\omega = i2\pi nT$ onto the real axis. As a result we obtain the following expression for the quantity $\sum_{\varepsilon} \mathcal{F}_{\varepsilon-\omega}$, which determines $\Delta_\omega(\boldsymbol{k})$ according to (4.29):

$$T\sum_{\varepsilon} \mathcal{F}^{+}_{\varepsilon,\,\varepsilon-\omega} \to -\frac{1}{4\pi i}\int\limits_{-\infty}^{\infty} d\varepsilon \left[\operatorname{th}\frac{\varepsilon}{2T}\mathcal{F}^{+A}_{\varepsilon,\,\varepsilon-\omega} - \operatorname{th}\frac{\varepsilon-\omega}{2T}\mathcal{F}^{+R}_{\varepsilon,\,\varepsilon-\omega}\right] -$$

$$-\frac{1}{4\pi i}\int\limits_{-\infty}^{\infty} d\varepsilon \int\limits_{-\infty}^{\infty} d\varepsilon_1 \sum_{\omega'}\Big\{[G^{+R}_{\varepsilon\varepsilon_1}\Delta^{*}_{\omega'}G^{-A}_{\varepsilon_1-\omega',\,\varepsilon-\omega} + \mathcal{F}^{+R}_{\varepsilon\varepsilon_1}\Delta_{\omega'}\mathcal{F}^{A}_{\varepsilon_1-\omega',\,\varepsilon-\omega}] +$$

$$+ [G^{+R}_{\varepsilon\varepsilon_1}\mathcal{F}^{A}_{\varepsilon_1-\omega',\,\varepsilon-\omega} + \mathcal{F}^{+R}_{\varepsilon\varepsilon_1}G^{-A}_{\varepsilon_1-\omega',\,\varepsilon-\omega}] \times$$

$$\times \left[\frac{e}{c}(\boldsymbol{v}\boldsymbol{A}_{\omega'}) - e\varphi_{\omega'}\right]\Big\}\left(\operatorname{th}\frac{\varepsilon_1}{2T} - \operatorname{th}\frac{\varepsilon_1-\omega'}{2T}\right) \tag{4.31}$$

($\boldsymbol{A}$, φ are electromagnetic potentials). This expression can be written

$$T\sum_{n>0}\mathcal{F}^{+}_{\varepsilon-\omega,\varepsilon} + T\sum_{n<0}\mathcal{F}^{+}_{\varepsilon,\varepsilon+\omega} + 4\pi i m p_0 e \tau_s (\Delta^{*}\psi)_\omega. \tag{4.32}$$

The first two terms in (4.32), which correspond to the first integral in (4.31), are not anomalous. They can be expanded, for example, in powers of Δ/T as in the steady-state case /327/.

The appearance of an anomalous term is highly significant. The third term in (4.32), which corresponds to the second integral on the right of (4.31), is anomalous. The function ψ is determined by comparing (4.31) and (4.32). The anomalous term involves integrals which contain products of advanced and retarded Green's functions in the coordinate representation.

In derivation of the steady-state Ginzburg-Landau equations /327/, the fact that the temperature Green's functions $\tilde{G}^0_\omega(R)$ decrease exponentially when $R > \xi_0$ was an important factor. The retarded and advanced Green's functions, however, do not decrease exponentially for large values of $t_1 - t_2$ but oscillate at frequency Δ. The structure of the power-series expansion of the anomalous terms is very complicated. The singularities which this expansion involves make it impossible simply to carry over the derivation of the steady state Ginzburg-Landau equations to the nonsteady-state case.

Alloys containing paramagnetic impurities are especially interesting, since at a certain impurity concentration gapless superconductivity arises /199/. Thanks to the absence of an energy gap, one can derive local equations for the time-dependent superconductivity parameter in the nonsteady-state case. That this is possible stems from the fact that when $\tau_s T \ll 1$ the Green's function decreases exponentially in time $t_1 - t_2 \sim \tau_s$ ($\tau_s^{-1} = \tau_1^{-1} - \tau_2^{-1}$ is the time between the collisions, during which reversal of electron spin takes place). Then

$$\operatorname{th}\frac{\varepsilon_1}{2T} - \operatorname{th}\frac{\varepsilon_1 - \omega'}{2T} \approx 2\omega'\delta(\varepsilon_1),$$

which substantially simplifies the expression for the anomalous part in (4.32).

It was shown in /328/ that the behavior of alloys with paramagnetic impurities is described by the following generalized Ginzburg-Landau equations:

$$\dot{\Delta} + \frac{\tau_s}{3}\left\{\pi^2 (T^2 - T_c^2) + \frac{|\Delta|^2}{2}\right\}\Delta + D\left(-i\nabla - \frac{2e}{c}A\right)^2\Delta = 0,$$

$$j = \sigma E - \frac{2\sigma\tau_s}{c}|\Delta|^2 (A - \nabla\chi), \quad D = \tau_1 v/3. \qquad (4.33)$$

The parameter $\varkappa \equiv \varkappa_{GL}$ in this generalized theory is equal to $\varkappa = (c^2/48\pi\sigma D)^{1/2}$. as can be seen from (4.33). Depending on what the initial pure metal is, the alloy may be either a Pippard ($\varkappa \ll 1$) or a London ($\varkappa \gg 1$) superconductor.

Equation (4.33) can be used to examine certain nonlinear effects in variable fields. We first mention the appearance of combination frequencies. If, for example, we consider a weak variable field $H = H_0 e^{i\omega t}$, applied parallel to the boundary of the half-space (the half-space $z < 0$ is assumed to be occupied by the superconductor), combined consideration of the first of equations (4.33) and Maxwell's equations leads to the following expression for $Q_{\omega+\omega'}$ ($Q = A - \nabla\chi$, $\Delta = \Delta' e^{2ie\chi/c}$):

$$Q_{\omega+\omega'} = \frac{16\pi\sigma\tau_s\Delta'^2}{c^2} \times$$

$$\times \frac{4e^2 D Q_\omega (Q^2)_{\omega'}}{c^2 (i\omega' - \tau_s\Delta'^2/2)\{[k_1(\omega') + k(\omega)]^2 + 4\pi\sigma i(\omega + \omega')/c^2 - \delta_L^{-2}\}} + Ae^{k(\omega+\omega')z}. \qquad (4.34)$$

The constant A is determined by the resonance boundary conditions.

In /328, 329/ the question of the destruction of superconductivity by high-frequency fields was also investigated. The situation turns out to be different for a superconducting half-space and for large sheets. In the first case superconductivity is preserved in a rapidly variable field and in the extreme case of high power. The superconductivity parameter can then be described by the equation (in dimensionless units; see (4.35))

$$\Delta(z) = \mathrm{th}(\varkappa z/2).$$

In sheets the situation is completely different. For example, for high frequencies, in sheets of thickness L such that $1 \ll L \ll \varkappa^{-1}$, superconductivity persists as long as the amplitude h of the field does not exceed a limiting value, which is $4\sqrt{2L\omega^{3/2}}$ (for $\varkappa \gg 1$).

Several significant points have come to light in investigation of thin films with high concentrations of nonmagnetic impurities /331/. The analysis was based on the equations obtained by Eliashberg in the case of a low concentration of paramagnetic impurities $\tau_s T_c \gg 1$ (the only condition assumed was $\tau_s\Delta \ll 1$ at temperatures sufficiently close to T_c, $(T_c - T)/T_c \ll 1/(\tau_s\Delta)^2$). The presence of a certain quantity of paramagnetic impurities results in some smoothing of the singularity in the spectrum.

In the dimensionless variables of the Ginzburg-Landau theory, the equations for the time-dependent parameter ψ in the one-dimensional case have the form

$$\begin{gathered}\left(-\frac{\partial}{\partial t} + \frac{1}{\varkappa^2}\frac{\partial^2}{\partial z^2}\right)\psi + (1 - A^2)\psi - \psi^3 + \psi U = 0,\\ \left(-\frac{\partial}{\partial t} + \frac{1}{\varkappa^2}\frac{\partial^2}{\partial z^2}\right)U = \nu\frac{\partial(\psi^2)}{\partial t},\\ -a\frac{\partial A}{\partial t} + \frac{\partial^2 A}{\partial z^2} - \psi^2 A = 0,\end{gathered} \tag{4.35}$$

where $\nu = \pi^3\tau_s T/14\zeta(3)$. The appearance in (4.35) of the anomalous term containing the function $U(z, t)$ is highly significant. Equations of a similar type without terms in $U(z, t)$ were discussed earlier in /333, 334/. The question of nonlinear equations for Δ in pure superconductors was considered in /335/.

In the case of a thin film, when ψ may be assumed constant across the thickness, and the external magnetic field $H = H_0 \sin\omega t$ parallel to the film is equal to the external field, equations (4.35) become

$$-\frac{1}{2}\frac{\partial\rho}{\partial t} + (1 - A^2)\rho - \rho^2 + U\rho = 0, \quad -\frac{\partial U}{\partial t} = \nu\frac{\partial\rho}{\partial t}. \tag{4.36}$$

We have denoted $\rho = \psi^2$. The solution of these equations for large times can be written

$$\rho(t) = \frac{\omega}{\nu+1}\left[1 - \exp\left\{-\frac{2\pi}{\omega}(1 + \nu\bar{\rho} - \overline{A^2})\right\}\right] \times \\ \times\left[\int_0^{2\pi} d\chi' \exp\left\{-\frac{1+\nu\bar{\rho} - \overline{A^2}}{\omega}\chi' - \frac{A^2}{\omega}[\sin\chi - \sin(\chi - \chi')]\right\}\right]^{-1}. \tag{4.37}$$

The condition for the existence of a steady state solution is $1+\nu\bar{\rho}-\bar{A}^2>0$. If this condition is not fulfilled, the superconductivity parameter decays exponentially after the field is applied.

A peculiar situation arises for frequencies and fields which satisfy the conditions $\omega \ll \overline{A^2}$, $\nu(1-\overline{A^2}) \ll \omega$. In this case

$$\rho(t)=(1-\overline{A^2})\,2\sqrt{\frac{\pi}{\omega}}\exp\left\{-\frac{1-\sin 2\omega t}{\omega}\right\}$$

is a strongly oscillating quantity.

In /330/ an adiabaticity criterion was investigated. Estimates for the characteristic frequency Ω_0 were obtained. At frequencies $\omega \ll \Omega_0$, the energy gap adiabatically follows the field. For small superconductors, whose properties in a stationary field were investigated previously by Larkin /336/, it turned out that the adiabatic pattern does not arise in the entire frequency range where $\omega\tau \gg 1$ (τ is the relaxation time).

In superconducting allows Ω_0 depends on the value of the Ginzburg-Landau parameter, whose determination has already been considered for these cases. For example, when $\varkappa \sim 1$, $\Omega_0 \sim \Delta_0^3/T_c^2$, and in the London limiting case ($\varkappa \gg 1$), $\Omega_0 \sim (1/\varkappa^2)(\Delta_0^3/T_c^2)$. Thus, the adiabatic pattern occurs only at very low frequencies.

At the opposite extreme, when $\omega \gg \Omega_0$ and, of course, the adiabatic situation is completely absent, the oscillating part of the energy gap is nevertheless small, so that one can use the averaged Ginzburg-Landau equations.

Of considerable interest is the frequency range $\Omega_0 \ll \omega \ll \Omega_1$, in which the penetration of the field into the superconductor is described by London's equation (Ω_1 is determined by the condition that when $\omega \gg \Omega_1$ the skin effect dominates). It is significant that the inequality $\Omega_1 \gg \Omega_0$ delineates the indicated frequency range.

In /330/ the breakdown of superconductivity in a suitable high frequency field is discussed. The equations

$$(\Delta^2-1)\Delta-\frac{1}{\varkappa^2}\frac{\partial^2\Delta}{\partial z^2}+\overline{A^2}(z)\Delta=0,\quad \partial^2 A/\partial z^2=\dot{A}+\Delta^2 A \qquad (4.38)$$

involve the time derivative of the field $\dot{A}$, corresponding to the electric field contribution to the current. If this term is neglected, equations (4.38) revert to the steady-state equations investigated by Ginzburg /337/. As shown in /330/, the corresponding curves exhibit a hysteresis effect. In the case under discussion, inclusion of the term $\dot{A}$ gives rise to a new branch with $\Delta^2 \sim \omega$ and a breakdown of superconductivity on this branch in a high frequency field. This breakdown disappears in very thin superconducting films of thickness L comparable to the penetration depth of the field.

References /338, 339/ are devoted to the nonlinear electrodynamics of superconducting films. In /338/ the critical power at which the superconducting state breaks down was calculated. It was shown that at thicknesses L such that $L \ll \delta_L$ (δ_L is the London penetration depth) the major factor in the breakdown of superconductivity is the electric component of the field.

Reference /338/ presents a highly significant result, pertaining to the dependence of impedance on power: this dependence possesses hysteresis. Specifically, the breakdown of the superconducting state and the transition into the normal phase occurs at a power W_1 significantly greater than the power W_2 at which the opposite transition into the superconducting state occurs when the field is decreased.

In /339/ it was shown that throughout the frequency range $\tau_e^{-1} \ll \omega \ll \Delta_0$ (τ_e is the homogeneous relaxation time) the dependence of the energy gap in the film on the field is essentially nonadiabatic. In the case

$$\nu \frac{3}{2\rho_0^2} \gg 1 \qquad \left(\rho_0 = \frac{1}{2\pi T} \frac{2De^2H_c^2d}{c^2}\right)$$

a detailed quantitative treatment of the hysteresis nature of the transmittance coefficient of UHF-radiation /338/ has been performed.

The dependence of the critical current on the quantity η ($\eta = \Delta(T)/\Delta_0$) has also been investigated. Reference /339/ employed the equations

$$\begin{gathered}\alpha\left[\left(1-\frac{H}{H_c}\right)-\frac{\langle\widetilde{A}_x z\rangle}{{}^1/_3 H_c d^2}-\frac{\widetilde{A}^2}{{}^2/_3 H_c^2 d^2}\right]\Delta- \\ -\beta\frac{\partial\Delta}{\partial t}-\gamma\Delta^3+2\pi TU\Delta=0, \qquad (4.39) \\ \boldsymbol{j}=\sigma\left[\boldsymbol{E}-\frac{1}{\pi Tc}\psi'\left(\frac{1}{2}+\rho_0\right)\widetilde{\boldsymbol{A}}\Delta^2\right],\end{gathered}$$

derived by Gor'kov and Eliashberg /340/ for the case of a superconductor in a strong magnetic field ($1 - H/H_c \ll 1$). Here $\widetilde{\boldsymbol{A}}$ is the vector potential of the variable field $d = L/2$; the field is assumed to be parallel to the xy-plane (the plane of the film); $\langle\ldots\rangle$ denotes averaging over the thickness of the film:

$$\alpha = 4\pi T\psi'\left(\frac{1}{2}+\rho_0\right)\rho_0;$$

$$\beta = \frac{1}{2}\psi'\left(\frac{1}{2}+\rho_0\right);$$

$$\gamma = \frac{x}{8\pi T}\left[-\psi''\left(\frac{1}{2}+\rho_0\right)-\frac{\rho_0}{3}\psi'''\left(\frac{1}{2}+\rho_0\right)\right];$$

$\psi(x)$ is the logarithmic derivative of the gamma-function. In a strong field, just as in the presence of paramagnetic impurities, there may be gapless superconductivity. This enables one to set up a local differential equation for Δ.

The nonlinear electrodynamics of pure superconductors ($l \gg \xi_0$) was investigated in /341/. In this case too, however, impurities play an essential role. Transition to the steady state demands allowance for the relaxation frequency $\omega_p \sim \tau_e^{-1}$. Even if the impurity concentration is very low and the value of τ_e is correspondingly very large, for sufficiently low field frequencies $\omega \sim \omega_p$ the relaxation mechanism must be taken into consideration. Moreover, the presence of impurities results in a widening of the surface levels /342/, so that perturbation theory may be applied, relative to a constant magnetic field. In /341/ the dependence of surface impedance on an external magnetic field was discussed. It was shown that fluctuations in the ordering parameter lead to several special features of the nonlinear increment to the surface impedance.

In the usual model, based on consideration of the four-fermion interaction, inelastic scattering processes are not allowed for. Therefore, this model does not describe, for example, relaxation processes in the system after application of the field is applied. In /332/ a general dynamic scheme was constructed, allowing for energetic relaxation mechanisms. Examples of such mechanisms are electron-phonon and electron-electron interactions. The basic equations of the phonon model were formulated by Eliashberg in /89/. In the usual superconductivity theory they form the basis for the theory of strong coupling superconductors (see Chapter I, Section 8). Equations furnishing a dynamic generalization of the phonon model were derived in /332/. These are the most general equations of the dynamic theory.

A high-frequency field may influence the value of Δ in different ways. For example, absorption of field energy by excitations is accompanied by changes in energy distribution. This in turn results in a change in Δ. For thin films at frequencies $\omega \gg (\gamma T_c)^{1/2}$ ($T_c - T \sim T_c$, $T \sim T_c$) and $\omega \gg (\gamma T_c)^{1/2} (1 - T/T_c)^{1/4}$ ($T_c - T \ll T_c$) this mechanism is the chief source of the change in Δ. In this case one can considerably simplify the general equations of the dynamic theory. It turns out that the average $\bar{\Delta}$ is determined by the same equation as in the BCS theory ($\gamma = \tau_0^{-1}$):

$$\Delta = \lambda \int_{-\omega_D}^{\omega_D} \frac{d\varepsilon \cdot \Delta}{\sqrt{\varepsilon^2 - \Delta^2}} (1 - 2n_\varepsilon),$$

but the excitation distribution function n_ε is not the equilibrium distribution function and is described by the kinetic equation

$$D\left(\frac{e}{c}\right)^2 A_\omega A_{-\omega}\{U_-(n_{\varepsilon-\omega}-n_\varepsilon)-U_+(n_\varepsilon-n_{\varepsilon+\omega})+ \\ +V(1-n_\varepsilon-n_{\omega-\varepsilon})\}=I^{ph}+I^e,$$

$$U_\pm=[\varepsilon(\varepsilon\pm\omega)+\Delta^2]\,\Theta(\varepsilon\pm\omega-\Delta)/\varepsilon\,[(\varepsilon\pm\omega)^2-\Delta^2]^{1/2},$$

$$V=[\varepsilon(\omega-\varepsilon)-\Delta^2]\,\Theta(\omega-\varepsilon-\Delta)/\varepsilon\,[(\omega-\varepsilon)^2-\Delta^2]^{1/2};$$

I^{ph} and I^e are the collision integrals of electron excitations with phonons and each other.

It was shown in /343, 344/ that a high frequency field whose frequency exceeds a certain critical value ω_c, determined by the equation

$$\omega_c^2\ln\frac{8\Delta}{\omega_c}=2\pi\tau_0^{-1}\Delta, \tag{4.40}$$

results in an increase in the ordering parameter Δ (τ_0 is the energy relaxation time of the excitations). There may thus be a superconducting state stimulated by the field. When $\omega>\omega_c$, the electron excitation distribution function $n(\varepsilon)$ is shifted toward the region of high energy values. The possible increase in the energy gap by a high frequency field is bounded by $\Delta(0)$.

*BIBLIOGRAPHY**

1. Bardeen, J., L. Cooper, and J. Schrieffer. — Phys. Rev. 108 (1957), 1175.
2. Bogolyubov, N.N. — ZhETF 34 (1958), 58; Bogolyubov, N.N., V.V. Tolmachev, and D.V. Shirkov. A New Method in the Theory of Superconductivity. — Izdatel'stvo AN SSSR, 1958 (Russian) [English translation: Consultants Bureau. New York, 1959].
3. Gor'kov, L.P. — ZhETF 34 (1958), 735.
4. Abrikosov, A.A. and I.M. Khalatnikov. — UFN 65 (1958), 551.
5. Bardeen, J. and J. Schrieffer. Recent Developments in Superconductivity. — In: Progress in Low Temperature Physics 3, Chap. VI, 1961.
6. Lynton, E.A. Superconductivity (2nd edition). — London, Chapman and Hall, 1969.
7. Blatt, J.M. Theory of Superconductivity. — New York, Academic Press. 1964.
8. Geilikman, B.T. and V.Z. Kresin. — UFN 99 (1969), 51.
9. Superconductivity. — "Nauka," 1967. (Russian)
10. Kulik, I.O. and I.K. Yanson. The Josephson Effect in Superconductive Tunneling Structures.— "Nauka," 1970 [English translation: IPST Cat. No.2231. Jerusalem, 1972].
11. Shoenberg, D. Superconductivity. — London, Cambridge University Press, 1952.
12. Gorter, C. and H. Casimir. — Phys. Zs. 35 (1934), 963.
13. Bethe, H. and A. Sommerfeld. Electron Theory of Metals. — Moscow, ONTI, 1938. [Russian translation]
14. Geilikman, B.T. — ZhETF 34 (1948), 1042.
15. Zavaritskii, N.V. — ZhETF 39 (1960), 1571.
16. Bardeen, J., G. Rickayzen, and L. Tewordt. — Phys. Rev. 113 (1959), 982.
17. Maki, K.— Progr. Theor. Phys. 31, (1964), 378.
18. Andreev, V.V. and V.V. Slezov. — FMM 17 (1964), 477.
19. Cubo, R., M. Yokota, and S. Nakajima. — J. Phys. Soc. Japan 12, (1957), 1203.
20. Geilikman, B.T. and V.Z. Kresin. — ZhETF 41 (1961), 1142.
21. Kresin, V.Z. — ZhETF 36 (1959), 1947.
22. Landau, L.D. and I.Ya. Pomeranchuk. — ZhETF 7 (1937), 180; Phys. Zs. Sowjet 10 (1936), 649; cf. Landau, L.D. Collected Works, Vol.1, p.208. "Nauka," 1969. (Russian)
23. Guenault, A. — In: Superconductivity Conference. Cambridge, 1959.
24. Jones, R. and A. Toxen. — Proc. Eleventh Int. Conf. Low Temperature Physics. Toronto, 1969; Phys. Rev. 120 (1960), 1167.
25. Kadanoff, L. and P. Martin. — Phys. Rev. 124 (1961), 670.
26. Tewordt, L. Phys. Rev. 127 (1962), 371; 128 (1962), 12; 129 (1963), 657.
27. Luttinger, J. — Phys. Rev. 136A (1964), 1481.
28. Kresin, V.Z. — In: Trudy XVII Vsesoyuznogo soveshchaniya po fizike nizkikh temperatur, Donetsk, 1972.
29. Vasudevan, R. and C. Sung. — Phys. Rev. 144 (1966), 237.
30. Anderson, P. — J. Phys. Chem. Solids 11 (1959), 26.
31. Geilikman, B.T. and V.Z. Kresin. — DAN SSSR 123 (1959), 259.
32. Geilikman, B.T. and V.Z. Kresin. — ZhETF 36 (1959), 959.
33. Sladek, R. — Phys. Rev. 97 (1955), 902.

* [For abbreviated names of Russian periodicals, see list at end of bibliography.]

34. Peierls, R.E. Quantum Theory of Solids. – Oxford University Press, 1955.
35. Ziman, J.M. Electrons and Phonons. – Oxford University Press, 1960.
36. Zavaritskii, N.V. – ZhETF 34 (1958), 1116.
37. Renald, G. and C. Reynolds. – Phys. Rev. B. 2 (1970), 3533.
38. Satterthwaite, C. – In: Superconductivity Conference. Cambridge, 1959.
39. Chaudhuri, K. – Phys. Chem. Sol. 23 (1962), 1341.
40. Chang, G., R. Jones, and A. Toxen. – IBM J. Res. Developm. 6 (1962), 112.
41. Guenault, A. – Proc. Roy. Soc. A 262 (1961), 420.
42. Mendelssohn, K. and C. Renton. – Proc. Roy. Soc. 230 (1955), 157.
43. Laredo, S. – Proc. Roy. Soc. 229A (1955), 473.
44. Graham, G. – Proc. Roy. Soc. A248 (1958), 522.
45. Mendelssohn, K. and J. Olsen. – Proc. Phys. Soc. 63A (1950), 2.
46. Zavaritskii, N.V. – ZhETF 39 (1960), 1191.
47. Rowell, P. – Proc. Roy. Soc. A254 (1960), 542.
48. Mendelssohn, K. and H. Montgomery. – Phil. Mag. 1 (1956), 718.
49. Mendelssohn, K. – Phys. Suppl. 24 (1958), 53.
50. Mendelssohn, K. and R. Rowell. – Cryogenics 2 (1961), 26.
51. Toxen, A., G. Chang, and R. Jones. – Phys. Rev. 126 (1962), 919.
52. Biondi, M. et al. – Rev. Mod. Phys., 30 (1958), 1109.
53. Radosewich, L. and W. Williams. – Phys. Rev. 188 (1969), 770.
54. Connolly, A. and K. Mendelssohn. – Proc. Roy. Soc. A266 (1962), 429; Souca, J. – Proc. Phys. Soc., s.2 (1969) 629.
55. Hulm, J. – Proc. Roy. Soc. 204A (1950), 98.
56. de Haas, W. and A. Rademakers. – Physica, 7 (1940), 992; Comm. Phys. Lab. Univ. Leiden, 254 (1940) 261e.
57. Shiffman, C. – Proc. Phys. Soc. 71 (1958), 597.
58. Morris, D. and M. Tinkham. – Phys. Rev. Lett. 6 (1961), 600.
59. Mendelssohn, K. and C. Shiffman. – Proc. Roy. Soc. A255 (1960), 199; Mendelssohn, K. – IBM J. Res. Developm. 1 (1962), 27.
60. Khalatnikov, I.M. – ZhETF 36 (1959), 1818.
61. Zavaritskii, N.V. – ZhETF 37 (1959), 1506; In: Materialy 5-go Vsesoyuznogo soveshchaniya po fizike nizkikh temperatur (1958), p.55. – Izdatel'stvo AN GSSR, 1960.
62. Scatterthwaite, C. – Phys. Rev. 125 (1962), 873.
63. McConnell, B. – Bull. Amer. Phys. Soc., s.II, 15 (1970) 821.
64. Lindefeld, P. and M. Moskowitz. – Proc. Twelfth Int. Conf. Low Temperature Physics, p.238, Kyuto Kaikan. Kyoto, 1970.
65. Mendelssohn, K. and J. Olsen. – Phys. Rev. 80 (1950), 859.
66. Webber, K. and D. Spohr. – Phys. Rev. 91 (1953), 414.
67. Hulm, J. – Phys. Rev. 90 (1953), 1116.
68. Detwiler, D. and H. Fairbank. – Phys. Rev. 88 (1952), 1049.
69. Zavaritskii, N.V. – ZhETF 38 (1960), 1573.
70. Strässler, S. and P. Wyder. – Phys. Rev. Lett. 10 (1963), 225.
71. Abrikosov, A.A. and N.V. Zavaritskii. Appendix to Russian translation of /11/. – IL, 1955. (Russian)
72. Laredo, S. and A. Pippard. – Proc. Cambr. Phil. Soc. 51 (1955), 368.
73. Andreev, A.F. – ZhETF 46 (1964), 1823.
74. Krylov, I.P. and Yu.V. Sharvin. – ZhETF, Pis'ma, 12 (1970), 102.
75. Landau, I.L. – ZhETF, Pis'ma, 11 (1970), 437.
76. Kapitsa, P.L. – ZhETF 11 (1941), 1.
77. Khalatnikov, I.M. – ZhETF 22 (1952), 687.

78. Lifshits, E.M. and Yu.V. Sharvin. – DAN SSSR 79 (1951), 783.
79. Shikin, V.B. – ZhETF 53 (1967), 1431.
80. Andreev, A.F. – ZhETF 47 (1964), 2222.
81. Watson, J. and G. Graham. – Canad. J. Phys. 41 (1963), 1738.
82. Geilikman, B.T. and V.Z. Kresin. – In: "Problemy mnogikh tel i fizika plazmy" (Trudy Mezhdunarodnogo simpoziuma, Novosibirsk, 1965), p.144. "Nauka," 1967.
83. Geilikman, B.T. and V.Z. Kresin. – FTT 7 (1965), 3294.
84. Geilikman, B.T. and V.Z. Kresin. – ZhETF, Pis'ma, 5 (1967), 271.
85. Geilikman, B.T. and V.Z. Kresin. – FTT, 9 (1967), 3111.
86. Scalapino, D. and K. Wada. – Phys. Rev. Lett. 14 (1965), 102; Swihart, J., Y. Wada, and D. Scalapino. – Phys. Rev. Lett. 14 (1965), 106.
87. Scalapino, D., J. Schrieffer, and J. Wilkins. – Phys. Rev. 148 (1966), 263.
88. Ambegaokar, V. and J. Woo. – Phys. Rev. **139A** (1965), 1818.
89. Eliashberg, G.M. – ZhETF **38** (1960), 966; **39** (1960), 1437.
90. Nambu, Y. – Phys. Rev. 117 (1960), 648.
91. Brockhouse, B. et al. – Phys. Rev. 128 (1962), 1099.
92. Zavaritskii, N.V. – ZhETF, Pis'ma, 5 (1967), 434; ZhETF 57 (1969), 752.
93. Chen, J. et al. – Phys. Lett. 25A (1967), 679.
94. Claeson, T. and G. Grimvall. – J. Phys. Chem. Sol. 29 (1968), 387.
95. Pokrovskii, B.L. – ZhETF 40 (1961), 641.
96. Geilikman, B.T., R.O. Zaitsev, and V.Z. Kresin. – Proc. Tenth Int. Conf. Low Temperature Physics, IIA, p.173. Moscow, 1967.
97. Geilikman, B.T. – J. of Low. Temper. Phys. 4 (1971), 189.
98. Pokrovskii, V.L. and M.S. Ryvkin. – ZhETF 43 (1962), 92.
99. Zavaritskii, N.V., E.S. Itskevich, and A.N. Voronovskii. – ZhETF, Pis'ma, 7 (1968), 271; ZhETF 60 (1971), 1408.
100. Abrikosov, A.A. and A.P. Gor'kov. – ZhETF **35** (1958), 1558; **36** (1959), 319.
101. Heer, C. and J. Daunt. – Phys. Rev. 76 (1949), 854.
102. Mendelssohn, K. Cryophysics. New York, Wiley (Interscience), 1960.
103. Ginzburg, V.L. – ZhETF 14 (1944), 177.
104. Ginzburg, V.L. Superconductivity. – Moscow, Izdatel'stvo AN SSSR, 1946. (Russian)
105. Kresin, V.Z. and V.A. Litovchenko. – ZhETF 53 (1967), 2154.
106. Pokrovskii, V.L. – ZhETF 40 (1961), 641.
107. Cubo, R., M. Yokota, and S. Nakajima. – J. Phys. Soc. Japan 12 (1957), 1203.
108. Ziman, J.M. Principles of the Theory of Solids. – London, Cambridge University Press, 1964.
109. Lifshits, I.M., M.Ya. Azbel', and M.I. Kaganov. Electron Theory of Metals. – "Nauka," 1971. (Russian)
110. Josephson, B. Private communication.
111. Dorfman, Ya.G. and I.K. Kikoin. Physics of Metals. – GTII, 1933. (Russian)
112. Geilikman, B.T. and V.Z. Kresin. – ZhETF 39 (1960), 502.
113. Ginzburg, V.L. – ZhETF 21 (1951), 979.
114. Gorter, C. – Can. J. Phys. 34 (1956), 1334.
115. Bömmel, H. – Phys. Rev. 96 (1954), 220; 100 (1955), 758.
116. Mackinnon, L. – Phys. Rev. 98 (1954), 1181; **100** (1955), 655.
117. Landau, L.D. – ZhETF 16 (1946), 574.
118. Privorotskii, I.A. – ZhETF 43 (1962), 1331.
119. Morse, R. and H. Bohm. – Phys. Rev. 108 (1957), 1094.
120. Morse, R. – Progr. Cryogenics 1 (1959), 220.
121. Morse, R. – IBM J. Res. Developm. 6 (1962), 58.

122. Pokrovskii, V.L. – ZhETF 40 (1961), 898.
123. Pokrovskii, V.L. and V.A. Toponogov. – ZhETF 40 (1961), 1112.
124. Privorotskii, I.A. – ZhETF 42 (1962), 450.
125. Ginzburg, V.L. – J. of Phys 8 (1944), 148.
126. Litovchenko, V.A. – FTT 6 (1964), 1051.
127. Geilikman, B.T. and V.Z. Kresin. – ZhETF 40 (1961), 970.
128. Geilikman, B.T. and V.Z. Kresin. – FTT 5 (1963), 3549.
129. Shepelev, A.G. – UFN 96 (1968), 217; UFN 98 (1969), 753.
130. Khokhenberg, P. – ZhETF 45 (1963), 1208.
131. Ambegaokar, V. and A. Griffin. – Phys. Pev. 137A (1965), 1151; Bennemann, K.H. – Phys. Lett. 14 (1965), 273.
132. Bezuglyi, P.A., A.A. Galkin, and A.P. Korolyuk. – ZhETF 36 (1959), 1951; 39 (1960), 7.
133. Morse, R., J. Olsen, and Y. Gavenda. – Phys. Rev. Lett. 3 (1959), 15, 193.
134. Shepelev, A.G. – ZhETF 45 (1963), 2076.
135. Shepelev, A.G. and G.D. Filimonov. – ZhETF 48 (1965), 1054.
136. Dobbs, E. and G. Thomas. – Proc. Tenth Int. Conf. Low Temperature Physics, IIB, p.260. Moscow, 1967.
137. Perz, J. and E. Dobbs. – Proc. Roy. Soc. A297 (1967), 408.
138. Claiborne, L. and N. Einspruch. – Phys. Rev. Lett. 15 (1965), 862; Proc. Tenth Int. Conf. Low Temperature Physics, IIB, p.256. Moscow, 1967.
139. Shepelev, A.G. and G.D. Filimonov. – ZhETF 56 (1969), 472.
140. Fil', V.D., O.A. Shevchenko, and P.A. Bezuglyi. – Proc. Tenth Int. Conf. Low Temperature Physics, IIB, p.265. Moscow, 1967.
141. Bohm, H. and N. Horwitz. – Proc. Eighth Int. Conf. Low Temperature Physics, p.191. London, 1963.
142. Shepelev, A.G. – ZhETF 45 (1963), 2076.
143. Claiborne, L. and N. Einspruch. – Phys. Rev. 151 (1966), 229.
144. Hart, H. and B. Roberts. – Bull. Am. Phys. Soc. 7 (1962), 175.
145. Saunders, G. and A. Lawson. – Phys. Rev. 135 (1964), A1161.
146. Weil, R. and A. Lawson. – Phys. Rev. 141 (1966), 452.
147. Jones, C. and J. Rayne. – Phys. Lett. 21 (1966), 510.
148. Fil', V.D., O.A. Shevchenko, and P.A. Bezuglyi. – ZhETF 52 (1967), 891.
149. Fossheim, K. and J. Leibowitz. – Phys. Lett. 22 (1966), 140.
150. Bliss, E. and J. Rayne. – Bull. Amer. Phys. Soc. 11 (1966), 241.
151. Dobbs, E. and J. Perz. – Proc. Eighth Int. Conf. Low Temperature Physics, p.195. London, 1963; Rev. Mod. Phys. 36 (1964), 257.
152. Shen, L., N. Senozan, and N. Phillips. – Phys. Rev. Lett. 14 (1965), 1025.
153. Akhiezer, A.I. – ZhETF 8 (1938), 1318, 1330.
154. Akhiezer, A.I., M.I. Kaganov, and G.Ya. Lyubarskii. – ZhETF 32 (1957), 837.
155. Pokrovskii, V.L. and S.K. Savvinykh. – ZhETF 43 (1962), 564.
156. Akao, F. – Phys. Lett. 30A (1969), 409.
157. Moskalenko, V.A. – FMM 8 (1959), 503.
158. Suhl, H., B. Matthias, and L. Walker. – Phys. Rev. Lett. 3 (1959), 552
159. Douglass, D. and L. Falicov. Progress in Low Temperature Physics 4 (1964), 97.
160. Geilikman, B.T., R.O. Zaitsev, and V.Z. Kresin. – FTT 9 (1967), 821.
161. Geilikman, B.T., R.O. Zaitsev, and V.Z. Kresin. – FMM 23 (1967), 796.
162. Geilikman, B.T., R.O. Zaitsev, and V.Z. Kresin. – FMM 25 (1968), 159.
163. Macvilar, M. and R. Rose. – J. Appl. Phys. 39 (1968), 1721; Phys. Lett. 26A (1968), 510.
164. Zavaritskii, N.V. – ZhETF 45 (1964), 1839; 48 (1965), 837.
165. Townsend, P. – Phys. Rev. 128 (1962), 591; Rev. Mod. Phys. 36 (1964), 207.

166. Cohen, R., B. Abeles, and G. Weisbarth. – Phys. Rev. Lett. 18 (1967), 336.
167. Garfunkel, M.– Rev. Mod. Phys. 36 (1964), 197.
168. Levy, M., F. Carcey, and R. Kaginawa. – Proc. Eleventh Int. Conf. Low Temperature Physics, Kyoto Kaikan. Kyoto, 1970.
169. Moskalenko, V.A. and M.E. Palistrant. – DAN SSSR 162 (1965), 539; ZhETF 49 (1965), 770; DAN SSSR 169 (1966), 812.
170. Moskalenko, V.A., L.Z. Kon, and M.E. Palistrant. – Proc. Tenth Int. Conf. Low Temperature Physics, IIA, p.181. Moscow, 1967.
171. van der Hoeven, B. Jr. and P. Keesom. – Phys. Rev. 137 (1965), 103.
172. Abrikosov, A.A., L.P. Gor'kov, and I.E. Dzyaloshinskii. Methods of Quantum Field Theory in Static Physics. – Gostekhizdat, 1962. (Russian)
173. Lacy, L. and A. Daniel. – Phys. Rev. Lett. 27 (1971), 1128.
174. Pippard, A.B.– Phil. Mag., 46 (1955), 1104.
175. Deaton, B.– Phys. Rev. Lett. 16 (1966), 577.
176. Fate, W. and R. Shaw. – Phys. Rev. Lett. 19 (1967), 230.
177. Randorff, J. and B. Marschall. – Phys. Rev. 2 (1970), 100.
178. Chopra, K. and T. Hutchinson. – Can. J. Phys. 36 (1959), 805.
179. Wu, H., H. Bohm, and R. Thomas. – Phys. Rev. 2 (1970), 122.
180. Newcomb, C. and R. Shaw. – Phys. Rev. 173 (1968), 509.
181. Perz, J. and E. Dobbs. – Proc. Roy. Soc. 113A (1967), 296.
182. Bliss, E. and J. Rayne. – Phys. Rev. 177 (1969), 673.
183. Kresin, V.Z. and Yu.V. Medvedev. – FMM 26 (1968), 182.
184. Woo, J. – Phys. Rev. 155 (1967), 429.
185. Pokrovskii, V.L. – ZhETF 40 (1961), 143.
186. Pokrovskii, V.L. and M.S. Ryvkin. – ZhETF 40 (1961), 1959.
187. Migdal, A.B.– ZhETF 34 (1958), 1438.
188. Bömmel, H. – Superconductivity Conference. Cambridge, 1959.
189. Dobbs, E. et al. – Proc. Twelfth Int. Conf. Low Temperature Physics, Kyoto Kaikan, p.119. Kyoto, 1970.
190. Kulik, I.O. – ZhETF 43 (1962), 1489.
191. Bardeen, J. and M. Stephen. – Phys. Rev. 136 (1964), 1485.
192. Ozaki, H. and N. Mikoshiba. – Phys. Lett. 23 (1966), 52.
193. Vas'kin, V.V. and B.Ya. Demikhovskii. – FTT 10 (1968), 419.
194. Chandrasekhar, B. and J. Rayne. – Phys. Rev. Lett. 6 (1961), 677.
195. Alers, C. and D. Waldorf. – Phys. Rev. Lett. 6 (1961), 677.
196. Kramer, E.J. and C.L. Bauer. – Phys. Rev. 163 (1967), 407.
197. Lee, B. and B. Deaton. – Phys. Rev. Lett. 23 (1969), 1438.
198. Deaton, B. – Phys. Lett. 32A (1970), 333.
199. Abrikosov, A.A. and L.P. Gor'kov. – ZhETF 39 (1960), 1781.
200. Kadanoff, L. and I. Falko. – Phys. Rev. 136, A (1964), 1170.
201. Gurevich, V.L.– ZhETF 38 (1960), 532.
202. Belozorov, A.D. and E.A. Kaner. – ZhETF 55 (1968), 642.
203. Andreev, A.F. and Yu.M. Bruk. – ZhETF 50 (1966), 1420.
204. Landau, L.D. and E.M. Lifshits. Electrodynamics of Continuous Media. – Gostekhizdat, 1957. (Russian)
205. Gottlieb, M. and M. Garbuny. – Phys. Lett. 24A (1967), 585.
206. Andreev, A.F. – ZhETF 53 (1967), 680.
207. Landau, L.D. – ZhETF 7 (1937), 371.
208. Lifshits, E.M. and Yu.V. Sharvin. – DAN SSSR 79 (1951), 783.
209. Gurevich, V.L. – ZhETF 37 (1959), 71, 1680.

210. Morse, R. and H. Bohm. – J. Acoust. Soc. Am. **31** (1959), 1523.
211. Morse, R. – Progr. Cryogenics **1** (1959), 220.
212. Morse, R. – IBM J. Res. Developm. **6** (1962), 58.
213. David, R., H. van der Laan, and N. Roulis. – Physica 29 (1963), 357.
214. David, R.– Philips Res. Rept. **19** (1964), 524.
215. Tsuneto, T.– Phys. Rev. **121** (1961), 402.
216. Kadanoff, L. and A. Pippard.– Proc. Roy. Soc. A **292** (1966), 299.
217. Cullen, J. and R. Ferrell.– Phys. Rev. 146 (1966), 282.
218. Levy, M., R. Kagiwada, and I. Rudnick. – Phys. Rev. 132 (1963), 2039.
219. Hart, H. and B. Roberts.– Bull. Am. Phys. Soc. **7** (1962), 175.
220. Levy, M.– Phys. Rev. 131 (1963), 1497.
221. Silin, V.P. and G.L. Kotkin.– FMM 14 (1962), 456.
222. Leibowitz, J.– Phys. Rev. 133A (1964), 84.
223. Perz, J.– Can. J. Phys. 44 (1966), 1765.
224. Leibowitz, J. and K. Fossheim.– Phys. Rev. Lett. 21 (1968), 1246.
225. Abrikosov, A.A.– ZhETF **32** (1957), 1442.
226. Caroli, C. and M. Cyrot.– Phys. Kondens. Materie **4** (1965), 285.
227. Maki, K.– Phys. Rev. **158** (1967), 397.
228. Lindefeld, P., E. Lynton, and R. Soulen.–Proc. Tenth Int. Conf. Low Temperature Physics, IIA, p.396. Moscow, 1967.
229. Mendelssohn, K.–IBM J. Res. Developm. **6** (1962), 27; Lowell, J. and K. Mendelssohn.– Proc. Tenth Int. Conf. Low Temperature Physics, IIA, p.402. Moscow, 1967.
230. Lowell, J. and J. Sousa.– Phys. Lett. A25 (1967), 469.
231. Muto, Y. et al.– J. Phys. Soc. Japan 23 (1967), 130; Proc. Tenth Int. Conf. Low Temperature Physics, IIA, p.407. Moscow, 1967.
232. Delveccio, L. and P. Lindefeld.– Phys. Rev. **B1** (1970), 1097.
233. Parks, R. and F. Zumsteg.– Phys. Rev. Lett. 18 (1967), 47.
234. Galaiko, V.P. and I.I. Fal'ko.– Proc. Tenth Int. Conf. Low Temperature Physics, IIA, p.424. Moscow, 1967; ZhETF 52 (1967), 976.
235. Geilikman, B.T. and V.Z. Kresin.– Phys. Lett. **40A** (1972), 123.
236. Maki, K.– Phys. Rev. **148** (1966), 370.
237. Maki, K.– Phys. Rev. 156 (1967), 437; Proc. Tenth Int. Conf. Low Temperature Physics, IIA, p.170. Moscow, 1967.
238. Cooper, L., A. Houghton, and H. Lee.– Phys. Rev. 148 (1966), 198.
239. Kagiwada, R. et al.– Phys. Rev. Lett. 18 (1967), 74.
240. Otter, F. and P. Solomon.– Phys. Rev. Lett. **16** (1966), 681; Proc. Tenth Int. Conf. Low Temperature Physics, IIA, p.328. Moscow, 1967.
241. Vijfeijken, A.– Phys. Lett. 23 (1966), 65; Proc. Tenth Int. Conf. Low Temperature Physics, IIA, p.316. Moscow, 1967.
242. Ohta, T.– Japan, J. Appl. Phys. **6** (1967), 645.
243. Fiory, A. and B. Serin.– Phys. Rev. Lett. **16** (1966), 308.
244. Fiory, A. and B. Serin.– Phys. Rev. Lett. **19** (1967), 227.
245. Huebener, R.– Phys. Lett. A25 (1967), 588.
246. Baixeras, J. and S. Williamson.– Solid State Comm. **5** (1967), 599.
247. Vijfejken, A. and A. Nillsen.– Philips Res. Rept. **20** (1965), 505.
248. Valatin, I.C.– Nuovo Cim. **7** (1958), 843.
249. Hebel, L.C. and C.P. Slichter.– Phys. Rev. 113 (1959), 1504; **107** (1957), 901; Hebel, L.C.– Phys. Rev. 116 (1959), 79.
250. Fibich, M.– Phys. Rev. Lett. 14 (1965), 561, 621.
251. Redfield, A.G. and A.G. Anderson.– Phys. Rev. 116 (1959), 583.

252. Masuda, Y. and A.G.Redfield.– Phys. Rev. 125 (1962), 159.
253. Redfield, A.G.– Phys. Rev. Lett. 3 (1959), 85.
254. Masuda, Y.– IBM J. Res. Developm. 6 (1962), 24.
255. Hammond, R.H. and W.D.Knight.– Phys. Rev. 120 (1960), 769.
256. Hammond, R.H. and G.M.Kelly.– Proc. Ninth Int. Conf. Low Temperature Physics, p.492. New York, 1965.
257. Masuda, Y.– Phys. Rev. 126 (1962), 1271.
258. Masuda, Y. and A.G.Redfield.– Phys. Rev. 133A (1964), 944; Reif, F. – Phys. Rev. 106 (1957), 208.
259. Hammon, R.H. and G.M.Kelly.– Rev. Mod. Phys. 36, No.1 (1964), 185; Androes, C.H. and W.D.Knight.– Phys. Rev. 121 (1961), 779; Phys. Rev. Lett. 2 (1959), 386.
260. Pincus, P. et al.– Phys. Lett. 13 (1964), 21; W.Fite II and A.G.Redfield.– Phys. Rev. Lett. 17 (1966), 17; Proc. Tenth Int. Conf. Low Temperature Physics, IIA, p.384. Moscow, 1967.
261. Delrieu, J.M. and J.M.Winter.– Solid State Comm. 4 (1966), 545.
262. Asayama, K. and Y.Masuda.– J. Phys. Soc. Japan, 21 (1966), 1459.
263. de Gennes, P. – Proc. 14 Coll. Ampere, North Holland Publish. Co., 1967.
264. Mattis, D.C. and J.Bardeen.– Phys. Rev. 111 (1958), 412.
265. Miller, P.B.– Phys. Rev. 118 (1960), 928.
266. Abrikosov, A.A., L.P.Gor'kov, and I.M.Khalatnikov.– ZhETF 35 (1958), 265; 37 (1959), 187.
267. Edwards, S.F.– Phil. Mag. 3 (1958), 1020.
268. Pokrovskii, V.L. and M.S.Ryvkin.– ZhETF 40 (1961), 1859.
269. Medvedev, Yu.V.– FMM 30 (1970), 235.
270. Maki, K.– Phys. Rev. Lett. 14 (1965), 98.
271. Ovchinnikov, Yu.N.– ZhETF 59 (1970), 128; 50 (1966), 1364.
272. Maki, K.– Progr. Theor. Phys. 31 (1964), 831.
273. Urushadze, G.I.– ZhETF 50 (1966), 404.
274. Caroli, C. and K.Maki.– Phys. Rev. 159 (1967), 316.
275. Maki, K.– Proc. Conf. on Supercond., Stanford, 1969.
276. Hibler, W.D. III and M.Cyrot.– Proc. Conf. on Supercond., Stanford, 1969.
277. Glover, R.E. III and M.Tinkham.– Phys. Rev. 108 (1957), 243.
278. Ginsberg, D.M. and M.Tinkham.– Phys. Rev. 118 (1960), 990.
279. Richards, P.L. and M.Tinkham.– Phys. Rev. 119 (1960), 575.
280. Norman, S.L. and D.H.Douglass.– Proc. Tenth Int. Conf. Low Temperature Physics, IIB, p.251. Moscow, 1967.
281. Palmer, L.H. and M.Tinkham.– Phys. Rev. 165 (1965), 588; Tinkham, M. – Proc. Tenth Int. Conf. Low Temperature Physics, IIB, p.238. Moscow, 1967.
282. Norman, S.L. and D.H.Douglass.– Phys. Rev. Lett. 18 (1967), 339.
283. Biondi, M.A. and M.P.Garfunkel.– Phys. Rev. Lett. 2 (1959), 143; Phys. Rev. 116 (1959), 853.
284. Biondi, M.A., M.P.Garfunkel, and A.O.McCoubrey.– Phys. Rev. 108 (1957), 495.
285. Thompson, W.A.– Phys. Lett. A24 (1967), 353.
286. Lehoczky, A. and C.V.Briscoe.– Phys. Rev. Lett. 23 (1969), 695; Proc. Conf. on Supercond., Stanford, 1969.
287. Schmidt, H.– Proc. Conf. on Supercond., Stanford, 1969; Zs. f. Phys. 216 (1968), 336; Mikeska, H.Y. and H.Schmidt.– Zs. f. Phys. 238 (1970), 239.
288. Khaikin, M.S.– ZhETF 34 (1958), 1389.
289. Pippard, A.B.– Proc. Roy. Soc. 192 (1947), 370, 339.
290. Kolpagiu, M.K. and L.Z.Kon.– Phys. Stat. Sol. 39 (1970), 61.

291. Kaplan, R., A.H. Nethercot Jr., and H. Boorse.– Phys. Rev. 116 (1959), 270.

292. Biondi, M.A., A.T. Forrester, and M.P. Garfunkel.– Phys. Rev. 108 (1957), 497.

293. Rudheimer, N.M. et al.– Proc. Ninth Int. Conf. Low Temperature Physics, p.381. New York, 1965.

294. Waldram, J.R.– Proc. Tenth Int. Conf. Low Temperature Physics, IIA, p.207. Moscow, 1967; Rev. Mod. Phys. 36 (1964), 187; Adv. in Phys. 13 (1964), 1.

295. Richards, P.L.– Phys. Rev. Lett. 7 (1961), 412.

296. Adkins, C.J.– Proc. Poy. Soc. A268 (1962), 276.

297. Lewis, R.T.– Phys. Rev. **134A** (1964), 1.

298. Biondi, M.A., M.P. Garfunkel, and W.A. Thompson.– Phys. Rev. **A136** (1964), 1471.

299. Zemon, S.A. and M.A. Boorse.– Phys. Rev. 146 (1966), 309.

300. Leslie, J.D. et al.– Phys. Rev. A134 (1964), 309.

301. Norman, S.L.– Phys. Rev. 167 (1968), 393.

302. Pokrovskii, V.L. and M.S. Ryvkin.– ZhETF 43 (1962), 900.

303. Ginsberg, D.M. and J.D. Leslie.– IBM J. Res. Developm. 6 (1962), 55; Rev. Mod. Phys. 36, No.1 (1964), 198; Phys. Rev. 133A (1964), 362.

304. Finnemore, D.K. and D.E. Mapother.– Phys. Rev. 140A (1965), 507.

305. Spiewak, M. – Phys. Rev. 113 (1959), 1479; Phys. Rev. Lett. 1, No.4 (1958), 136.

306. Budzinski, W.V. and M.P. Garfunkel.– Phys. Rev. Lett. 16 (1966), 1100; 17 (1966), 24.

307. White, R.H. and M. Tinkham.– Phys. Rev. A136 (1964), 203.

308. Hibler, W.D. III and B.W. Maxfield.– Phys. Rev. Lett. 21 (1968), 742.

309. Gittleman, J.I. and B. Rosenblum.– Phys. Rev. Lett. 16 (1966) 734; Proc. Tenth Int. Conf. Low Temperature Physics, IIA, 380. Moscow, 1967.

310. Shepelev, A.G., O.P. Ledenev, and G.D. Filimonov.– ZhETF, Pis'ma, 14 (1971), 428.

311. Trivisonno, J. et al.– J. of Low Temper. Phys. 4 (1971) 97.

312. Gurevich, V.L.– ZhETF 37 (1959), 71.

313. Blount, E.J.– Phys. Rev. 114 (1959), 418.

314. Silin, V.P.– ZhETF 38 (1960), 977.

315. Kontorovich, V.M.– ZhETF 45 (1963), 1638.

316. Kaner, E.A. and V.G. Skobov.– ZhETF 45 (1963), 610.

317. Melik-Barkhundarov, T.K.– ZhETF 55 (1968), 657.

318. Geilikman, B.T.– In: Trudy XVII Vsesoyuznogo soveshchaniya po fizike nizkikh temperatur. Donetsk, 1972.

319. Geilikman, B.T.– J. of Low Temper. Phys. 9 (1972), 553.

320. Moskalenko, V.A. – FMM 23 (1967), 585.

321. Fal'ko, I.I.– FMM 23 (1967), 391; FTT 9 (1967), 2124.

322. Aslamazov, L.G. and A.I. Larkin.– FTT 10 (1968), 1104.

323. Moskalenko, V.A., A.A. Golub, and L.Z. Kon.– FMM 30 (1970), 289.

324. Gor'kov, L.P. and M.A. Fedorov.– ZhETF, Pis'ma, 14 (1971), 568.

325. Larkin, A.I. and Yu.N. Ovchinnikov.– ZhETF 61 (1971), 2147.

326. Galkin, A.A., V.M. Svistunov, et al. – ZhETF 59 (1970), 77.

327. Gor'kov, L.P.– ZhETF 36 (1959), 1918.

328. Gor'kov, L.P. and G.M. Eliashberg.– ZhETF 54 (1968), 612.

329. Gor'kov, L.P. and G.M. Eliashberg.– ZhETF 54 (1968), 2430.

330. Gor'kov, L.P. and G.M. Eliashberg.– ZhETF, Pis'ma, 8 (1968), 329; ZhETF 56 (1969), 1927.

331. Eliashberg, G.M.– ZhETF 55 (1968), 2443.

332. Eliashberg, G.M.– ZhETF 61 (1971), 1254.

333. Abrahams, E. and T. Tsuneto.– Phys. Rev. 152 (1966), 416.

334. Caroli, C. and K. Maki.– Phys. Rev. 164 (1967), 591.

335. Kemoklidze, M.P. and L.P. Pitaevskii.– ZhETF 50 (1966), 243; 52 (1967), 1556.
336. Larkin, A.I.– ZhETF 48 (1965), 232.
337. Ginzburg, V.L.– ZhETF 34 (1958), 113.
338. Kulik, I.O.– ZhETF 57 (1969), 600.
339. Gor'kov, L.P. and N.B. Kopnin.– ZhETF 59 (1970), 234.
340. Gor'kov, L.P. and G.M. Eliashberg.– J. of Low Temp. Phys. 2 (1970), 161.
341. Fedorov, M.A.– ZhETF 59 (1970), 1716.
342. Pincus, P.– Phys. Rev. 158 (1967), 346.
343. Eliashberg, G.M.– ZhETF, Pis'ma, 11 (1970), 186.
344. Ivlev, B.I. and G.M. Eliashberg.– ZhETF, Pis'ma, 13 (1971), 464.

ABBREVIATIONS

DAN SSSR – Doklady Akademii Nauk SSSR
FMM – Fizika Metallov i Metallovedenie
FTT – Fizika Tverdogo Tela
UFN – Uspekhi Fizicheskikh Nauk
ZhETF – Zhurnal Eksperimental'noi i Teoreticheskoi Fiziki

SUBJECT INDEX